CANAL 250

THE STORY OF BRITAIN'S CANALS

CANAL 250

THE STORY OF BRITAIN'S CANALS

ANTHONY BURTON

The
History
Press

First published 2011

The History Press
The Mill, Brimscombe Port
Stroud, Gloucestershire, GL5 2QG
www.thehistorypress.co.uk

© Anthony Burton, 2011

The right of Anthony Burton to be identified as the Author
of this work has been asserted in accordance with the
Copyrights, Designs and Patents Act 1988.

British Library Cataloguing in Publication Data.
A catalogue record for this book is available from the British Library.

ISBN 978 0 7524 5952 3

Typesetting and origination by The History Press
Printed in Great Britain
Manufacturing managed by Jellyfish Print Solutions Ltd

CONTENTS

INTRODUCTION

I N OLDER history books the year 1760 is notable as the year George III came to the throne. In more modern books it has a far greater significance than just adding another I to the list of Georges. It is the date that is conventionally taken as marking the start of the Industrial Revolution. An essential part of that great social and technological upheaval was the transport route that was to bring in the raw materials and carry out the finished products of the new industries – the canals. In 1760, the Act of Parliament was passed that authorised the construction of the Bridgewater Canal, and that had a very special significance. It was the first canal in Britain that was independent of any natural waterway; and marked the beginning of the Canal Age, which saw a watery web stretch from Cornwall to the Highlands of Scotland. This book is a celebration of that achievement.

Anyone who writes about canals today owes a debt to those who went before. My own interest in canals was sparked by nothing more than going on holiday, but I was fascinated by the structures I saw along the way and, in particular, by the awe-inspiring Pontcysyllte aqueduct. I wanted to know more, and in particular I wanted to know about the man who was credited with building it, Thomas Telford. I read L.T.C. Rolt's biography, which gave a vivid picture of the life of the great engineer. It was the first time that I had even considered that engineering could be as much about personalities as practicalities. It was a revelation. I also acquired Eric de Maré's *The Canals of England*. What impressed me about that book was the way in which it showed the rich texture and fascinating contrasts of the waterways. In particular there were pictures of the lock staircase at Grindley Brook, which I had passed through on my journey along the Llangollen. At the time, I had been too busy trying to work out how to operate three interconnected locks to worry about much else. These photographs opened my eyes to the way in which the visual pleasures of the locks was dependent on details such as the contrast between the massive stones of the lock walls and the pattern created by the brick ridges on the ramp. In de Maré's own words: 'they achieve something supreme in utilitarian beauty'. I would certainly not quarrel with that. Shortly afterwards, I began to take a keener interest in the history of canals and inevitably had to turn to the work of Charles Hadfield. No one had ever attempted to write a comprehensive history of the canals before that and, even now, I can still marvel at the huge effort involved in his pioneering work. These three authors represent three strands of the story: the human story in Rolt, the physical story in de Maré and the detailed documentary history in Hadfield. The three approaches may be different, but each is valid and valuable and each has been influential in my own work.

Over the years I have spent countless hours in archives and I have always thought of this part of the work as a treasure hunt. You spend a vast amount of time digging around and occasionally you strike gold. Most official documents can be dry stuff, but there are nuggets in there that make the whole search worthwhile. There are even some real discoveries to be made. Years ago, working in what was then the British Transport Historical Record Office at Paddington, I ordered up one of those files you open more in hope than expectation. It was simply labelled 'Miscellaneous', but inside I found handwritten letters from Thomas Telford, which seemed never to have been catalogued. There is nothing quite like holding original letters to give you a sense of connection with the past. While researching in libraries and archives I was helped by the work already put in by Charles Hadfield, which made my own life so much easier – and was very conscious that he had set a high standard of meticulous research. But I was also always on the lookout for the little detail that brought the human story alive, in the way that Tom Rolt has done so magnificently in so many books. I remember coming across an entry in one of the Committee books of the Leeds and Liverpool Canal Company, recording the death of a workman in a fall of earth. He was simply described as 'a stranger called Thomas Jones supposed from Shropshire'. The phrase seemed to sum up something very important about the life of the navvy. To his employers he was no more than one unit in an anonymous army. I wanted to look behind the anonymity and find out more about these men, whose Herculean labours created the canals.

The story of canals is not just to be found in old records and documents; it is also written on the landscape. Much as I have enjoyed archive research, I have to confess that I have had even more pleasure from seeing the canals themselves. I have always contended that there is no real substitute to travelling them by boat. That way, you not only see how things work, you experience them directly. But you also have to learn how to look and understand the significance of what you see. That was something I had learned from Eric de Maré.

This book comes after decades, during which I have travelled all over Britain looking at canals; boating those that are navigable and tramping the countryside, hunting out the remains of some that are all but forgotten. In everything I have done, however, I have always been conscious of the work of those three great pioneers who first enthused me and taught me how to see and investigate the canal heritage. They remain an inspiration and without their pioneering efforts this survey of 250 years of canal history would never have been written.

The author would like to thank British Waterways for providing funding for the illustrations for this book, and the following for supplying the pictures (illustrations not listed are from the author's collection):

Illustrations on page 16, National Railways Museum; 19, 25, 34, 38, 39, 41, 45, 52, 54, 57, 59, 66, 68, 69, 72, 73, 78, 80, 84 (Bottom two), 86, 89, 91, 93, 97 (Top), 98, 100, 103, 107, 111, 114, 118, 120, 123, 125, 128 (Bottom), 134, 138, 141, 149, 150, 153, 160 (Bottom), 171 Waterways Trust; 22 (Left), 29, City Museum, Stoke-on-Trent; 22 (Right), Josiah Wedgwood & Sons Ltd; 24, National Portrait Gallery; 56, A.D. Cameron; 166 (Top), 168, Cotswold Canals Trust; 174, British Waterways.

1

THE FATHER OF CANALS

FRANCIS EGERTON, third Duke of Bridgewater, was a most unlikely candidate for the role as leader of a transport revolution. Nothing in his early life suggested or hinted at what was to come. It is true that the Egerton family had a fine tradition of public service behind them. His great-great grandfather, Sir Thomas Egerton, had been Lord Keeper of the Seal during Elizabeth I's reign; his great grandfather John Egerton, the first Earl of Bridgewater, had been a famous patron of the arts, and his own father was the first Duke. Francis was born in 1736, one of five sickly boys, four of whom died young. His early life was made even worse by the death of his father in 1745. His mother remarried within the year, to a man half her age, and seems to have taken little interest in his welfare – his new stepfather even less. He was considered to be rather stupid, if not almost imbecilic, and scarcely worth bothering about. His elder brother had inherited the title and Francis could be left to his own devices, since nothing was expected of him. All that changed in 1748, when his elder brother died, and he was no longer merely young Francis: he was the third Duke of Bridgewater. A younger son could be safely ignored, but not a Duke. His parents' first thoughts were to declare him unfit to hold the title, and they refused to pay for his education. It took a court case to settle that. Inevitably, the boy was deeply unhappy, ran away from home, and did what many teenagers have done down the years in similar circumstances. He started drinking. Something had to be done. He would need an education to prepare him for his place in the aristocratic world. There was a well-established method available for the young man – he must go on The Grand Tour, accompanied by a personal tutor. There he would learn refinement and manners and pick up an appropriate smattering of classical culture.

The young Duke's mentor was no ordinary teacher. Robert Wood was one of the foremost archaeologists of the day, among the first to apply a scientific approach to excavation and measurement of sites. He had spent a good deal of time working among the Hellenic and Roman remains of Palmyra and probably had little enthusiasm for trailing around Europe with a bored, drunken teenager. But he needed the money. So the scholar and the Duke set off to tour the usual sites, starting in Paris, and Wood's worst fears were soon realised. His young charge continued his drinking and pursued a French actress. It was all very scandalous. Wood whisked the Duke off to safer ground, visiting the sites of the ancient world and encouraged his young student to purchase a number of antiquities along the way, which were eventually crated up and sent off to the family home in Lancashire. It is an indication of the Duke's enthusiasm that, it is said, he never bothered to unpack them. At least he picked up enough sophistication and knowledge to make him presentable for an appearance in the English Court.

The famous portrait of the young Duke of Bridgewater, pointing to the Barton aqueduct.

The case for canals: one horse pulling a boat could carry the entire load of this team of pack-horses.

The entrance to the mines at Worsley.

There is a famous portrait of the Duke taken just a few years later, when he was in his twenties, that shows a very handsome and elegant young man, fashionably dressed and almost foppish. He was, at this time, very much the man about town, enjoying horse racing and gambling, and he must have had an air about him, for he successfully wooed one of the great beauties of the day. Elizabeth Gunning and her sister Maria were born into a very impoverished but aristocratic Irish family. Their mother thought seriously of putting them on the stage, but decided instead that they might do rather better for themselves in the more profitable marriage market. They appeared at the Court in borrowed clothes, but their exceptional good looks more than compensated for the lack of fortune. They were painted by Reynolds and had soon acquired husbands: Maria married the Earl of Coventry and Elizabeth became Duchess of Hamilton. Maria's marriage was short lived. Her husband died and it was as a widow that she met and was courted by the young Duke of Bridgewater in 1758. The Duchess and her sister were everything that the young man was not, sophisticated and worldly. But it was one thing for a young buck to have his fling – a future Duchess was supposed to behave very differently. He was now alarmed to hear rumours about the immoral behaviour of the Countess of Coventry. He promptly demanded that her sister should agree never to see her again as a condition of his marriage. Maria, not surprisingly, refused this rather ludicrous request. The match was over. The Duke left London in 1758 and renounced the company of women for ever. It was not just a case of not getting involved in any more romantic entanglements: he refused even to allow female servants in his house. He set off for his estates at Worsley, near Manchester, and threw himself into the industrial world.

It seems somewhat odd to us today to think of aristocrats engaging actively in industry. But the mid-eighteenth century saw the first stirrings of the Industrial Revolution and it was clear that one commodity would be increasingly profitable. The burgeoning iron industry was already using coke instead of charcoal in its furnaces. In a rural economy, there was wood for the fires in the surrounding countryside, but the growing towns needed a very different source of fuel for heating and cooking. Everyone needed coal. Many noble families began to realise that what lay deep under their grounds might be worth more than anything that might grow on the surface. The Duke of Bridgewater was one of the lucky ones: he had his own mines just up the road from his home in Worsley. There was just one problem: how to get the coal from the mines to Manchester. This was a town that was growing at a phenomenal rate. In 1717, the population had been around 8,000, now in 1757 it had risen to around 20,000. These would be his customers and they were rather less than 8 miles away.

This might not seem much of a difficulty: put the coal in carts, harness up the horses and away you go down the road. The road, however, was the problem not the solution. In the middle of the eighteenth century roads varied between bad and wretched. Under the old system of highway maintenance, the local people were charged with keeping the roads in good order. Citizens were supposed to give up a certain amount of time a year to repairing damage. In practice, those who could afford to do so, paid others to go in their place, while the unpaid did as little as possible. Improvement was supposed to arrive with the turnpikes. These were superior and travellers had to pay tolls to use them. The trouble was they were seldom much better than the older versions. The Turnpike Trustees generally farmed out the work of maintenance to whoever put in the lowest bid and sat back to enjoy the profits. Lowest prices generally went with shirked responsibilities. Eventually, things deteriorated to the point where the Trustees found themselves bound to act. A contemporary described the process: 'A meeting is called, the Farmer sent for and reprimanded, and a few Loads of Gravel buried among the Mud, serve to keep the Way barely passable.'

Arthur Young, who travelled around Britain reporting on the state of agriculture in the 1760s, had this to say about a Lancashire road:

> Let me most seriously caution all travellers, who may accidentally purpose to travel this ter-
> rible country, to avoid it as they would the devil; for a thousand to one but they break their
> necks or their limbs by overthrows or breaking downs. They will here meet with ruts which
> I actually measured four feet deep, and floating with mud only from a wet summer; what
> therefore must it be after a winter!

And this was the new Wigan turnpike. It is not hard to imagine what it must have meant for a team of horses trying to haul a loaded wagon through such a morass. Things looked very different, however, if the same load could be carried in a boat. Engineers did tests later in the century and found that the maximum load that a horse could pull on a wagon over what was euphemistically described as 'a soft road' was little more than half a ton. But if they set that same horse to pull a barge on a river, it could move 30 tons.

The actual figures may not have been available to an earlier generation, but it was a matter of simple observation that river transport was quicker and more economical than road trans-port. As a result, the rivers of Britain had been steadily improved for centuries. The biggest change came with the introduction of what were known as 'pound locks', though later we dropped the pound part and simply called them locks. The idea is simple. Natural rivers have an unfortunate tendency to behave in a very irregular fashion – sometimes they wander along placidly, then they will suddenly make a bit of a dash downhill in rapids and shallows. To overcome that problem, the river engineers arranged for boats to travel up and down in a series of controlled steps. Weirs were built across the river to create deep water. An artificial cutting was then constructed to bypass the weir, with a lock to overcome the difference in levels. The very first wholly modern lock in Britain was built on the River Lea in the six-teenth century, an event that was considered so important that it was described in a poem of 1577, 'Tale of the Two Swannes'. It may not be great poetry, but at least it tells us something about the technology:

The Canal du Midi inspired
the Duke: the impressive
lock staircase at Fonserannes.

… the locke
Through which the boates of Ware doe passe with malt,
This lock contains two double doors of wood,
Within the same a cesterne all of Plancke,
Which onely fils when boates come there to passe
By opening of these mightie dores.

This is clearly a timber-sided lock, with double doors. These must have been mitred doors, meeting at an angle, otherwise there would have been serious leaks. It seems an obvious thing, but they had to be invented in the first place and the man responsible was no less a person than Leonardo da Vinci. He may be famous as one of the world's greatest artists, but in 1482 he was the official engineer to the Duke of Milan. It was during this period of his life that he introduced the mitred gates for locks on the Naviglio Interno.

By the time the Duke of Bridgewater came to think about transport, river improvement had gone just about as far as it could go, but that still left areas that were some distance from the nearest navigable waterway. Unfortunately for him, Worsley was one of them. But if the Duke had forgotten everything he had learned about classical ruins since returning from his European jaunt, there was one memory that remained with him. He had visited the Languedoc Canal, now known as the Canal du Midi, in France.

This extraordinary canal was designed and constructed under the direction of Pierre-Paul Riquet. He was not an engineer, but a man who had the highly lucrative job of collecting the salt tax. The canal was designed to link the Atlantic and the Mediterranean, but there was one immediate problem that had to be solved. The canal had to cross a watershed, which meant that it had a summit from which water would drain down in two directions. If it was not constantly supplied with water it would simply dry up. Riquet found the highest point along the route and searched for the nearest water supply. He found it miles away in the Montagne Noir region, so his first task, when work began in 1666, was to build reservoirs and join them to his proposed canal by water channels, the rigoles. The first of these, the Rigole de la Montagne ran for 25km and was a major undertaking, but it was as nothing compared with the works on the canal itself.

This was a canal intended to take seagoing vessels, so the locks had to be big, with chambers 30m long and 5.6m wide. There were ninety-two of them altogether, and to save water and time for the boatmen, many were grouped together into staircases. In these, the top gates of the lower lock also form the bottom gates of the one above. Riquet built twelve double locks, four triples, one quadruple and it all reached a stupendous climax in the magnificent seven-lock staircase at Fonserannes. He crossed one river on an aqueduct and at Malpas – the 'bad step' – he took his canal through the hill in a tunnel. Almost a century before the Duke of Bridgewater began to think about waterways, Riquet had solved nearly all the major problems that future canal engineers could expect to face.

Here was the answer to the Duke's dilemma: he could build a canal to take the coal from his mines to Manchester. Among the Duke's most trusted employees were two brothers, Thomas and John Gilbert. Thomas was the Duke's Agent and his brother took a keen interest in the mines. We shall never know who came up with the first idea of canal building, but it was undoubtedly John Gilbert who had the largest say in the original plans. Most mines suffer from a problem with water that needs to be drained constantly to keep the workings dry. Gilbert's ingenious idea was to turn the problem into a solution. He devised a system in which the unwanted water would feed a system of underground canals that would reach to the coal faces and eventually emerge as a navigable waterway, heading off to Manchester.

To build a canal required an Act of Parliament and the Duke put in his application. It was approved in 1759, but with so many qualifications as to make it of comparatively little value. He could get as far as the River Irwell at Barton, and to continue his watery journey he could have made a connection with the river for the rest of the way. But the river authorities were having none of that. They had a valuable monopoly on trade and did not welcome interlopers. If the Duke wanted to use the river, he would have to pay large tolls – which rather defeated the whole object. He could not use the river and ensure his profits and it represented a barrier to further progress. At this stage another important character appears on the stage. Gilbert had been very impressed with a neighbour of his, a Derbyshire millwright called James Brindley, who had shown great expertise in managing water, when he had been called in to drain the aptly named Wet Earth Colliery. He was now recruited as canal engineer. Together, the Duke, Gilbert and Brindley came up with an altogether more ambitious scheme. If they could not join the Irwell, then they would vault over it. They would carry the Bridgewater Canal across the river on an imposing aqueduct. A new Act was passed in 1760 and now work could begin on the first canal in mainland Britain that took a line independent of any natural waterway.

Barton aqueduct; the print emphasises the monumental abutments.

The Act attracted very little attention at the time, but it marked the beginning of over half a century of canal construction that was to see the network develop right across the face of Britain, in the greatest period of civil engineering since the Romans built their roads.

It was one thing to get the Act of Parliament, but quite another to find the money to build a canal. This was a private venture and the Duke had to find the funds himself. He was a wealthy man but even he lacked the resources for such an enterprise. The first thing to go was the London house – he had no intention of ever again venturing into London society. Then he mortgaged his estates and after that he was forced to borrow money, first from relatives, then from Child's Bank. Even then, he was constantly short of funds. On one occasion, he was pursued by the local parson trying to get payment for a debt and was embarrassingly tracked down, hiding in a hay loft. It has to be remembered that he was still in his twenties, and there must have been many who condemned him as a flighty young nobleman, who was letting ambitions exceed the bounds of plain common sense. There was no shortage of so-called experts to denounce his folly. Although aqueducts were not new – the Romans had famously built them for water supply and Riquet had already made navigable aqueducts in France – there were still sceptics who doubted if such a scheme could succeed. An allegedly expert engineer, called in by the Duke to offer a second opinion, simply remarked that he had heard of castles in the air, but this was the first time he had heard of one being built. It is to the great credit of the Duke that he ignored the pessimist and went ahead with his plan.

There has always been doubt and controversy over who should take the credit for designing and building the great aqueduct over the river at Barton. There are arguments in favour of both

Although it was not recognised as such at the time, the Sankey Brook was arguably Britain's first true canal.

Brindley and Gilbert as being the leading figure, but as there is no doubt both were deeply involved in the project it is probably as well to let them share the credit. It is, however, notable that Brindley never attempted anything quite so bold again, but then he never again had to build an aqueduct over a navigable river. The structure was carried 38ft above the river on three arches that had to be high and wide enough to let the river craft pass safely underneath. Bridges of these dimensions were not unusual, but a bridge and an aqueduct are not the same. The latter, like the canal itself, has to be made watertight. This involved 'puddling', pounding a suitably dense clay with water until all the air had been forced out, leaving a uniform mass of material that could be spread out as a waterproof lining. The usual tool for pounding the mixture was the heavy boot of a workman – and stomping up and down all day in this cloying mess must have been exhausting.

When the great day came to fill the aqueduct with water, it seemed that the sceptics might be right after all. One of the arches began to give way and the water was quickly drained off again. Brindley did what he often did in times of stress – he took himself off to bed, which was not particularly helpful, leaving Gilbert to solve the problem. It was soon obvious that Brindley, who was naturally cautious, had been overly so in this case, piling on the clay puddle to a great depth. It was the sheer weight of the lining that was causing the problem. Gilbert got his men to remove tons of the puddle and on 17 July 1761 the water was let back in. This time there was no problem and the first barge, carrying 50 tons of coal, was hauled across. It was a triumph!

The aqueduct was the sensation of the day, drawing scores of tourists, who came to marvel at this 'canal in the air'. One visitor almost succumbed to a fit of the vapours at the sight:

> Whilst I was surveying it with a mixture of wonder and delight, four barges passed me in the space of about three minutes, two of them being chained together, and dragged by two horses, who went on the terras of the canal, whereon, I must own, I durst hardly venture to walk, as I almost trembled to behold the large River Irwell underneath me.

Georgian tourists were notoriously given to exaggeration; surely the poor, fainting gentleman must have seen a bridge crossing a large river. The quote does, however, give a hint of the excitement caused by this venture. Accounts appeared, not just in the local press, but also in London publications such as the *Gentleman's Magazine*. Other observers took a much more practical view of things:

> The difference in favour of canal navigation was never more exemplified nor appeared to full and striking advantage than at Barton-bridge in Lancashire, where one may see, at the same time, seven or eight stout fellows labouring like slaves to drag a boat slowly up the river Irwell, and one horse or mule, or sometimes two men at most, drawing five or six of the duke's barges, linked together, at a great rate upon the canal.

To be fair, this is a very partisan account by John Phillips, a propagandist for canals, and rather overlooks the fact that had the Irwell barge been going downstream, life would have been much easier for the boat crew. The real contrast was not with the river, but with the roads and here it was even more striking.

The canal was not, however, quite the pioneering marvel that people seemed to think. Ignoring what had happened in France – and the British were happy to ignore anything foreign – there had been canals built in Ireland and an important scheme had begun not far away

A sketch by Arthur Young, the travel writer, shows one of the small aqueducts on the Bridgewater Canal and also gives a good idea of the wretched nature of the road it passes over.

in 1755. The Sankey Brook Navigation was a canal in all but name. The Act was nominally 'for making navigable the River or Brook called Sankey Brook … in the county Palatine of Lancaster'. It attracted little attention, simply because it was seen as just another river improvement scheme. In practice, it consisted entirely of an artificial canal, and should by rights be applauded as the true beginner of the canal age. It would have been had anyone taken any notice. But its promoters realised that, if they had advertised the fact that they were doing something really radical, they might have attracted opposition. So they hid their intentions away by using the familiar language of river improvement. The Sankey Brook crept up on the world unnoticed. The Bridgewater Canal appeared with a dramatic flourish. It was the castle in the air that got the world talking.

The canal was working and the aqueduct proclaimed loud and clear that natural obstacles need not hamper the spread of canals. What very few noticed was that the aqueduct was only one part of a complex system. There was far more canal underground than appeared on the surface. In time, the system of underground waterways in the mine was to stretch for nearly 50 miles, working at many different levels. An American traveller, Samuel Curwen, visited Worsley in 1777 and was taken on a tour of the underground workings:

> We stepped into the boat, passing into an archway partly of brick and partly cut through the stone, of about three and a half feet high; we received at entering six lighted candles. This archway, called a funnel, runs into the body of the mountain almost in a direct line three thousand feet, its medium depth beneath the surface about eighty feet; we were half an hour passing that distance. Here begins the first under-ground road to the pits, ascending into the wagon road, so-called, about four feet above the water, being a highway for the wagons, containing about a ton weight of the form of a mill-hopper, running on wheels, to convey the coals to the barge or boats.

Special boats had been devised to reach the coal face. They were long and narrow, just 4ft 6in wide, and known as starvationers, because they were very skinny and their ribs showed. There was an equally ingenious system at the Manchester end of the canal, at Castlefield wharf. The coal had to be lifted to a higher level, so Brindley devised a system where the boats could be floated under a shaft. A crane, worked by a waterwheel, then lifted the coal to street level.

The whole canal was efficient and that efficiency was reflected in cheap carriage costs, which halved the price of coal in Manchester. Now the Duke was ready for an even more ambitious plan, to extend his canal to reach the Mersey at Runcorn. This would provide a link between two towns that were both developing with the growth of industry: Liverpool, the rapidly expanding port and Manchester, which was destined to become Cottonopolis, the heart of the cotton industry. There was intense opposition from the river authorities, but the Duke received his Act. However, difficulties in construction were not so easily overcome.

River crossings were simple compared with the problems faced at Barton. The land was flat, the rivers carried no traffic, so plain low structures were all that was needed. Then the engineers met an area of peaty bog, known as Sale Moor. As fast as the men tried to dig the canal, the mud simply oozed back in again. The problem was solved by digging trenches, and then lining them with timber baulks which were then fastened together to form parallel walls to keep the mud at bay, while the canal bed itself was excavated between them. The other problem was the crossing of the valley of the River Bollin, which called for a massive earth embankment. Although many came to wonder at Barton aqueduct, in strictly engineering terms building a high, stable embankment represented a far greater challenge. The final obstacle came with the arrival of the canal at Runcorn. Up to then it had been a broad, lock-free waterway, but now a flight of locks

An unusual feature of the Bridgewater was an early form of containerisation – boats carrying boxes designed to fit snugly into the hull.

had to be built. It was essential that they could take the Mersey flats, the traditional barges of the river, and these were 72ft long and 14ft 9in beam. It was a huge effort and Brindley was not to live long enough to see it completed.

The Duke was fortunate in one respect, that he could use his own workers and employees in the task. Samuel Smiles, in his *Lives of the Engineers*, written over a century after the work was completed, gives one of the very few accounts of who these men were and what they were like:

> In Lancashire proper names seem to have been little used at that time. 'Black David' was one of the foremen employed on difficult matters, and 'Bill o Toms' and 'Busick Jack' seem also to have been confident workmen in their respective departments. We are informed by a gentleman of the neighbourhood that most of the labourers employed were of a superior class and some of them were 'wise' or 'cunning men', blood-stoppers, herb-doctors, and planet-rulers, such as are still to be found in the neighbourhood of Manchester. Their very superstitions, says our informant, made them thinkers and calculators. The foreman bricklayer, for instance, as his son used afterwards to relate, always 'ruled the planets to find out the lucky day on which to commence an important work', and, he added, 'none of our work ever gave way'.

The Bridgewater canal system was an engineering triumph, but more importantly, as far as possible future investors in canals were concerned, it was a commercial success as well. Accounts for the later years of the canal must have made very satisfactory reading for the Duke who as a young man had risked so much. The income from the mines and the canal eventually amounted to almost £70,000 a year, the equivalent of several millions today. It is always difficult to interpret figures from another age, so to put that in some sort of perspective, it is worth looking at events that were to lead eventually to a news story that appeared in 2008. It is a vivid demonstration of the purchasing power of a gentleman on an income of £70,000 per annum.

Philippe, Duc d'Orleans, the nephew of Louis XIV, had assembled a magnificent collection of artistic masterpieces. In the 1790s his life, like that of every French aristocrat, was turned upside down by the Revolution. In order to try and build a new life, he decided to send his whole collection, of 305 paintings, for sale in London. The problem was that London was full of French aristocrats selling off their treasures: there were a glut on the market. The Duke of Bridgewater may have known little about art but he knew what he liked – he loved a bargain. He got together with his nephew Lord Gower and the Marquis of Carlisle, and together they bought the whole lot for £43,000. They'd managed to purchase over 300 Old Masters for less than the annual income from the Duke's industrial interests. And that's not the end of the story. The three then picked out the best of the bunch to keep for themselves. They selected ninety-four works and put the rest back up for auction; they fetched £42,500. So what did the Duke get himself for, in effect, his share of the outstanding £500? He finished up with some rare prizes: three Raphaels, four Titians, a Rembrandt self-portrait and eight Poussins. Some bargain! The story now moves forward in time.

Because the Duke died childless, the Bridgewater Collection went to the descendants of his sister, who had married the Duke of Sutherland. In later years, important paintings were sent on loan to the National Gallery of Scotland. In 2008 the present Duke of Sutherland decided to sell two of the most famous works, a pair of Titians, and offered them to the Gallery at what in the art market is considered the bargain price of £1 million pounds. Even given the huge value placed on Old Masters today, it is clear that the Duke's annual income made him a very rich man indeed.

In his later years, the Duke changed from the lithe young man dressed in style to one who was 'large and unwieldy, and seemed careless of his dress, which was uniformly a suit of brown'. He looked so little like an aristocrat that on one of his visits to Castlefield wharf, a workman called him over and asked for a hand loading coal into his barrow. The Duke duly obliged. He was a compassionate man: 'During the winter of 1774 the Duke of Bridgewater ordered coals to be sold to the poor of Liverpool in pennyworths, at the same rate as by cart loads. Twenty-four pounds of coal was sold for a penny.' His canals remained the great love of his life. On one of his rare visits to London, his gardener took the opportunity to beautify the grounds, planting extensive flowerbeds outside the windows. He did not get the response he expected. The Duke was livid; the blooms obstructed the view of the canal. He took his stick and demolished the lot: flowers came a very poor second to waterways.

The year 1760 has been traditionally taken as the starting point of the Industrial Revolution, and the start of work on the canal at Worsley is one of the key events. No doubt canals would have been built anyway; they were obviously of immense value to the developing industrial world. But someone had to take the first, bold step – and when that someone was prepared to risk his entire fortune on the venture, then he deserves to be known as 'The Father of Canals'. He died in 1803, after a road accident, by which time he had had the satisfaction of seeing his beloved canals spread throughout the land. He was buried among his ancestors at Little Gaddesden, and his memorial reads: 'He sent barges across fields the farmer formerly tilled'; not very romantic perhaps, but one feels that the Duke would have heartily approved.

2

THE BRINDLEY YEARS

I T IS difficult to appreciate the impact made by the opening of what now seems a modest canal from Worsley to Manchester, but at the time it was a sensation, and nowhere was its impact felt more strongly than in the newly developing industrial world. Britain was at the start of a period of rapid transformation. Work that had once been done by hand was now being performed by machines. Craftsmen working in their own homes or in small workshops were being replaced by wage earners in factories and mills. Villages were growing into towns; towns would become cities. The new world needed a new transport system, and it had found just what it needed: canals. It also needed experienced engineers to oversee their construction. It was inevitable that canal promoters would turn to the men who had built the Bridgewater Canal, but the Gilbert brothers had very successful careers working for the Duke of Bridgewater and saw no reason to change. That just left one man: James Brindley. James Sims, author of the *Mining Almanack* for 1849, looking back on this period, wrote, 'Amongst all the heroes and all the statesmen that have ever yet existed none have ever accomplished anything of such vast importance to the world in general as have been realised by a few simple mechanics.' It was just such men as James Brindley that he had in mind.

Brindley was born at Tunstead near Buxton in Derbyshire in 1716, but later moved to a farm near Leek in Staffordshire. We know very little about his early years. Samuel Smiles, who wrote biographies of many of the leading engineers of the eighteenth century, described Brindley's father as being a dissolute character, who did little for his family. But we do know that the boy must have received a rudimentary education, as some of his notebooks have survived. His spelling seems to have been entirely phonetic, and you can hear his local accent in the way he wrote. While working on an engine, he began with 'Bad louk', which improved to 'Midlin louk' and ended with the triumphant 'Engon at woork'. At the age of seventeen he was bound as apprentice to a wheelwright and millwright, Abraham Bennet of Sutton near Macclesfield.

Millwrights were the foremost mechanical engineers of the day, expected to turn their hands to the design and construction of many different types of machinery. There was another aspect of the work that was to have a vital impact on Brindley's later career: the millwright had to understand how to manage watercourses. In theory, the apprenticeship was a system in which the young man would learn the trade under the careful supervision of the master. Unfortunately, in this case, the millwright was often drunk and the journeyman was usually away. Not surprisingly, Brindley was regarded as something of a bungling incompetent, simply because no one had actually taken the time to teach him. So he set about a process of self-education, watching and copying others and learning from his own mistakes. It seems that at the very beginning of his career he had decided that if he wanted to know how to do something well, then he would

James Brindley posing with his surveying instrument.

The eminent potter Josiah Wedgwood, the chief promoter and first treasurer of the Trent & Mersey.

be better off relying on his own resources and common sense rather than alleged experts. It was an attitude that was to stay with him throughout his life.

The benefits of all his hard work appeared when he was first employed on his own as a millwright for a small silk mill near Macclesfield. The bungler proved that he had mastered his trade. It is difficult to disentangle myth from fact when looking at the early life of a man destined to become famous. John Phillips in his book *Inland Navigation*, written in 1805, idolised Brindley as the first, great canal engineer. He recounted this story of his early days. Bennet was commissioned to build a paper mill – something he had never attempted before and word got about that he was making a hash of it. So Brindley took it upon himself to put things right. He set off on a Saturday morning to walk 50 miles to the nearest existing paper mill. There he inspected all the machinery, memorised the details and walked home again, ready for work on Monday morning. After that he was able to ensure that work on the new mill was carried out to everyone's great satisfaction. It is a good story, but highly implausible. Even if you accept that anyone could walk a hundred miles in a weekend, and still find time to work out the technical details of a complex mill, it still doesn't ring true. The working machinery of the mill would have consisted of heavy hammers, powered by a waterwheel, which pulverised the rags used in paper making. Such giant hammers were commonplace and had been in use for centuries, in the fulling mills of the textile industry and in forges. No millwright could have been unaware of what was needed. What is clear, when the exaggerations have been swept away, is that Brindley was perfectly capable of rescuing the incompetence of the increasingly drunken Bennet. At least his master had enough sense to retire to his bottle and leave the young man to get on with running the business. When Bennet died in 1742 Brindley set up on his own, in Leek.

He soon established himself as being something more than a mere millwright. By now a new source of power had appeared in the land – the steam engine. The early engines, based on a design by Thomas Newcomen, were exclusively used for pumping, mainly from mines. Brindley went to see one of these new engines at Wolverhampton in 1756 and typically decided to build one himself, but using his own ideas. Bizarrely, he chose to construct it with a wooden cylinder and was sufficiently pleased with the result to patent it. No one ever repeated the experiment. But his growing reputation as an ingenious worker meant that people with problems were often knocking on his door. One such problem occurred at the Wet Earth Colliery near Manchester. The problem is contained in the name – water, and how to get rid of it. Brindley's solution was to install a pump, powered by a waterwheel, but he needed the water to power it. He had to take it from the Irwell, down an 800-yard channel to a point opposite the adit that gave access to the mine. Unfortunately, it was only possible to construct the channel on the opposite bank of the river to the mine, so he took the water underneath the river in an inverted siphon. Now it could be fed to the waterwheel to work the pumps. It was an immediate success – the mine was drained and became a highly profitable concern. It had an even more important consequence for Brindley: it brought his name to the attention of the Gilberts and launched him on his career as a canal engineer. But before that happened he had already made another important connection.

In 1750, he set up a second business at Burslem, in the heart of the Potteries, leasing the premises from the Wedgwood family. John and Thomas Wedgwood were, like other Staffordshire potters, starting to use powdered flint in their ware. The traditional earthenware of the district was made from a dark clay, that could either be disguised under a heavy white glaze or decorated by slip, creating patterns using a semi-liquid clay, rather like adding icing to a cake. It had been discovered that by adding flint to the clay before firing, a far paler product could be obtained. But in order to get the powder, the flints had first to be calcined, heated to a high temperature in a furnace, and then crushed in a mill. This was a potentially profitable business for a millwright. It was, however, to be one of the next generation of Wedgwoods who was to transform the industry.

Josiah Wedgwood was born in 1730, but a conventional childhood ended abruptly when his father died in 1739. Josiah's schooldays were over and he was apprenticed to his brother. An even more dramatic change came in 1742, when he fell victim to smallpox, a disease that was often fatal. Wedgwood recovered, but it left him with a permanently damaged leg, which eventually had to be amputated. If such a story of misery can be said to have a bright side, it was that the invalid had time to go back to his studies during his long convalescence. A regular visitor to the sickbed was his brother-in-law, the Revd William Willett. He encouraged Josiah to study science, and the more he studied the more it became clear to him that applying scientific ideas to manufacturing pottery represented the future. When he recovered he went to work with William Whieldon, a potter who produced ware far more sophisticated than was usual in Staffordshire. This gave Wedgwood just the opportunity he wanted to begin experimenting with new materials and techniques. In 1759, he began the first of his experimental notebooks and at the beginning he described his aims. The pottery of the region was now considered old-fashioned and crude. The taste was all for imported porcelain, not homely earthenware. In his own words, 'something new was wanted, to give a high spirit to the business'. It was time to try and match production to the demands of the customers. He was ready to start out on his own.

Wedgwood began to produce creamware, a light pottery using flints and delicate glazes. He worked systematically at improving the technology, but was also faced by another, rather different problem: the workforce. Like many manufacturing industries, work was still organised on craft principles, in which a master potter would make an entire piece himself. Wedgwood began to look at a very different way of working – breaking the making of a pot down into its different parts. One man would throw the pot, the next apply the handles, another the decoration and so on: he was developing a basic production line. To make it efficient, however, he needed to have everything under one roof, under his direct control. He needed a factory.

As Wedgwood developed his new ideas, it became increasingly clear that he needed a better transport system than the local roads. His flints came, for example, from East Anglia. He was to import clay from Devon and Cornwall. And he wanted to give his new, delicate pottery a smoother ride to market than would be possible on the back of a packhorse. He needed a canal. By 1765, his fame had spread far beyond the immediate neighbourhood – he had completed a dinner service for the Queen. Creamware was relabelled: from now on it would be Queensware. Josiah Wedgwood had become one of the most famous potters in the land so it was only natural that, when plans for a canal began to be formed, he should take the lead in its promotion. It was equally inevitable that he should turn to the man who had worked with the Wedgwoods in the past and had shown his canal construction abilities, James Brindley.

Canal promotion was about to enter a new and decisive phase. The Duke of Bridgewater had been able to finance the canal himself, but this was an altogether more ambitious scheme. The proposed canal was the Trent & Mersey, which became known as the Grand Trunk Canal, emphasising its importance. The estimate for construction was £130,000, which was to be raised by selling shares at £200 each. This was an age when the banking system was, at best, rudimentary, and the Company Treasurer was in effect banker to the enterprise. The role fell to Wedgwood, and it is a mark of how far he had come that he was able to provide a surety of £10,000, while the family as a whole bought £6,000 in shares. He was in it for the benefits the canal would bring to his industry, not for any potential profit on his shares. He certainly wasn't going to make any money in the short term, as he ruefully noted in a letter, where he listed the Company's officers:

James Brindley Surveyor General	£250 per ann.
Hugh Henshall Clerk of the Works	£150 per ann. For self & clerk
T. Sparrow Clerk to the Proprietors	£100 per ann.
Jos. Wedgwood Treasurer at £000 per ann. out of which he bears his own expenses.	

Before any work could begin, the all-important Act of Parliament had to be obtained. The Mersey and Irwell Navigation Company saw the proposal as a direct threat, and began issuing pamphlets, deploring the enterprise. They began by claiming that if the proposal went ahead it would 'involve two effectual Navigations, established under the Faith of Parliament, in certain and irreparable Ruin'. They then argued that the whole scheme was actually totally worthless and the canal was only being built 'to make a more lucrative job for engineers'. They did not seem to be aware of a glaring contradiction in their arguments. If the canal was entirely useless, how could it threaten them? Wedgwood rallied his supporters to produce pamphlets in favour of the canal, calling in his business partner Thomas Bentley and the eccentric genius Erasmus Darwin as authors. Unfortunately, they rather lost sight of the urgency of the situation and began quibbling about style. Wedgwood soon lost patience:

Must the Uniting of Seas & different countries depend upon the choice of phrase or mono-syllable? Away with such hypercriticism, & let the press go on, a Pamphlet we must have, or our design will be defeated, so make the best of the present, & correct, refine & sublimate in the next edition.

The work was rapidly finished. There was one other essential element to the application to Parliament – a detailed plan of the route and an estimate of building costs. The latter often

The arrival of the canal was a key factor in the development of the Potteries; in this photograph the canal itself is seen against a background of bottle ovens.

proved to have only a faint resemblance to the final figures – an aspect of public building works that has not changed greatly since then. Perhaps there is a case for bringing back the rules for public buildings that existed in ancient Rome. If the architect brought the building in below costs he was heaped with public honours; if it came within 25 per cent of the original estimate that was acceptable and the difference was made up out of the public purse. Anything above that, and the architect had to find the money himself. But before any estimates could be made, Brindley had to decide the route the canal would take and the engineering features that would be needed.

Brindley's first job was to survey the possible routes for the waterway. A canal has to be built on the level, with changes in height being made through the watery steps of locks. His habit was to ride over the land on horseback to get a feel for the terrain, after which the main work of taking levels could begin. Today he would have the advantage of an Ordnance Survey map, showing the contours of the land, but at that time the Ordnance Survey was not even in existence, so everything had to be done from scratch. The surveyors' tools were comparatively simple. Distances were measured by and in chains – the chain being both a physical object and a distance of 22 yards – which, incidentally, is how the cricket pitch came to acquire its seemingly arbitrary length. The principal levelling instrument consisted of a telescope mounted on a stand containing a spirit level. We tend to think of heights on maps in terms of heights above sea level, but this was not necessary for the canal surveyor. All he needed to know was the difference in levels along his proposed route, and he could establish any benchmark from which to measure the ups and downs. These were essential for finding differences in height. The method of working over short distances was to use tall poles, marked off in feet and inches. All the surveyor had to do was compare the reading at the distant pole with the one at telescope level. The telescope could be moved over a graduated horizontal scale to find angles, and directions were found using the simple magnetic compass. Mapping was to become far more sophisticated by the end of the century, when theodolites were being developed, but the simpler instruments of Brindley's day were more than adequate for the job. The maps of the routes that eventually emerged were themselves quite simple: hills for example were simply sketched in as if drawn by a child, with no indication of height. They did what they had to do: showed the Canal Committee and the Parliamentary Committee where the canal was going to go. Detail would be added later.

The main line of the Trent & Mersey is over 93 miles long, so inevitably there were going to be hills and valleys along the way. Brindley's preferred approach was to avoid obstacles rather than confront them; he would always go round a hill rather than cut through it. He followed the natural contours of the land as far as possible, and if that meant following a very serpentine course, then that was a small price to pay in order to avoid the expense of major construction works. This can be seen very clearly on another of his canals, the Staffs & Worcester near Kidderminster, where the artificial canal echoes the twists of the natural River Stour. There is a story that goes with this. One of the old canal myths describes how Brindley wanted to take his canal to join the Severn at the thriving river port of Bewdley, but the locals turned him down scorning his 'stinking ditch'. In reality the citizens petitioned the engineer to bring the canal to them, but that would have involved heavy engineering works, so he stayed with the easy line down the river valley. There, where river and canal met, a new inland port developed – Stourport. It thrived and the trade of Bewdley steadily diminished.

Brindley followed this technique of contour cutting throughout his working life, and the results can be seen most dramatically on the Oxford Canal north of Banbury. At Wormleighton, the canal almost circles the low hill, like a moat. Travellers who come this way by boat find themselves passing the front door of the house on the hill and about twenty minutes later they are looking at the back door of the same house. It was all very well for Brindley, but it became a source of huge annoyance to generations of boatmen. It was said that, on the northern Oxford, you could travel all day and still hear the same clock striking the hour on Brinklow church.

There was, however, one serious obstacle to progress on the Trent & Mersey. Harecastle Hill lay right across the line of the canal. It was all but impossible to cope by building locks up one side and down the other – apart from the inconvenience it would have been difficult to provide the water to keep the summit supplied. It was equally difficult to go round it, which left only one practical option: he would have to go through it in a tunnel. This was entering new territory. The first canal tunnel had been at Malpas on the Canal du Midi, but that was just 161m long and had been cut through soft rock. Brindley had no idea what he would meet once he started burrowing into the hill, but he did know how far he had to go. The tunnel would have to be almost 3,000 yards, well over 1½ miles long – and to add to his worries there would be three other shorter tunnels on the line. This was a problem that called for careful thought.

The other decision to be taken was on the locks. There were to be seventy-six in all, and he had to decide how to big to make them. On the Bridgewater Canal, the decision had been obvious from the start. To make the canal as effective as possible the locks were designed to take the barges already in use on the river. The Grand Trunk was to connect with the Bridgewater at its northern end and to the River Trent at the other, both of which carried broad-beamed barges. Logic suggested that the new canal should offer the same facilities. But there was a problem: if he made locks to accept vessels 14ft wide, then his tunnels would have to be wide enough to take them too. It was a daunting prospect. Brindley seldom asked for advice, but simply took himself off to a darkened room and there he lay until he had sorted things out for himself. He took the important decision: he could not build a tunnel to the appropriate dimensions. It has to be remembered that it was not just a question of doubling the amount of material that had to be excavated. The tunnel would need to be built in the form of a double

The Oxford Canal at Wormleighton beautifully demonstrates Brindley's preference for following the natural contours of the land.

The Dudley Canal was one of the first generation of new canals, and like the Bridgewater built to serve an underground complex of mines and quarries.

The narrow boat evolved as a result of decisions on lock size taken by Brindley. This heavily laden vessel is on the Coventry Canal.

arch – something close to a circle. It is not obvious when travelling on canals that there is an inverted arch below the waterline. Doubling the width would mean quadrupling the area of the cross-section – he would have to excavate four times as much material. It was simply too much for such a pioneering work. He decided to halve the width of the tunnels. Now they would only be able to take vessels approximately 7ft wide. The tunnel made no difference to the length of the boats, so he could keep lock length to the 70ft of the earlier canal. But if no boat wider than 7ft could go through the tunnel, then no boat wider than that could use the canals, so why bother to go to the expense of building wide locks? His plan called for a narrow canal to be worked by narrow boats.

The Act for the Trent & Mersey was duly obtained in 1766 and in the same year the Staffordshire & Worcestershire Canal was also approved, joining the Potteries to another great river, the Severn. This was to be followed by three more major works: the Oxford, the Coventry and the Birmingham canals, which would link together England's four great rivers – Thames, Severn, Trent and Mersey. It was an integrated system, so all were to be built with narrow locks. It set a pattern for English canals that was to have far-reaching effects. The new narrow boats were limited to cargoes of roughly 25 tons, considerably less than the river barges, but that was not important – the two systems were not competitors. The only competition was with the roads, and the narrow boats were a vast improvement on anything that could be offered by a horse and cart. It is easy for us to see that this was to prove a major handicap in the future, but at the time it seemed a compromise well worth making. For industrialists such as Wedgwood, the Trent & Mersey was everything he could have hoped for. He was planning a great expansion of his business and he chose a site for his new factory right alongside the canal. It was called Etruria, because he was setting out to capture the brilliance of the pottery of the ancient world that he believed was Etruscan. In fact, they weren't Etruscan at all, but he used the name not just for the works themselves, but also for the grand new family home that he built on the hill overlooking the canal, Etruria Hall. The original pamphlet had praised the notion of a canal as a picturesque feature in the landscape: 'to have a lawn terminated by water, with objects passing and repassing upon it, is a finishing of all other the most desirable'. Wedgwood envisaged a graceful curve to the canal as it swept past the new house, but Brindley the 'inflexible vandal' would have none of it. Wedgwood pleaded for his 'line of grace', but the engineer couldn't be budged from taking what he considered the best line. Wedgwood was, in any case, already facing criticism from other local potters, who claimed that the canal had been deviated to pass the new works, not appreciating that it was the works that had been sited to fit the canal. In time, Etruria was to develop into a small village and be home to one of the greatest of all English potteries.

The demands on Brindley grew steadily – by the 1770s he was chief engineer for ten canals and being consulted on others. The strain was enormous. And his habit of self-reliance did not help. Some of the decisions he took in those years seem strange and bizarre. One of the oddest examples can be seen at The Bratch on the Staffs & Worcester. Here a difference of levels of 30ft called for three locks to be built very close together. The obvious solution would have been to build a staircase of three interconnected locks. Instead he built the three separately with only a few feet of water between each. This is thoroughly inconvenient, as there is not enough room for boats to wait between the locks, so a boat going up has to wait until a boat going down has passed through all three before setting off itself. It is also puzzling – where does the water go when a lock is emptied? The answer can be found in ponds, mini-reservoirs – tucked away to the side of the canal. One thing can be said about the Staffs & Worcester is that it was one of the few canals, with which he was concerned, that was actually completed during his lifetime, on time and sufficiently close to being on budget as to have passed the Roman test. The other major project begun the same year is quite a different matter.

Wedgwood's own house, Etruria Hall, built to overlook the new canal.

The great obstacle was Harecastle tunnel. Brindley had reduced the dimensions as far as he could, not even allowing for a towpath. The boatmen had to leg their vessels through, lying on their backs and walking their feet along the walls of the tunnel. As there was no space for boats to pass, the tunnel was effectively one way – northbound traffic had half a day and southbound the other half. Early reports of construction were full of optimism. A visitor in 1767 wrote that the tunnel was the eighth wonder of the world and that Brindley 'handles rocks as easily as you would plumb-pies'. Others were more sceptical. When Brindley told a meeting that the whole project could be finished in five years, he was challenged and offered a wager that it could not be done in the time: 'Mr. B told him, that he never refused upon anything which he seriously asserted & offer'd them to article in a Wager of £200 that he perform'd what he had said.' He lost his bet, but never had to pay; Brindley died in 1772 and the tunnel was only eventually opened in 1777.

The work of tunnelling began with the initial survey, which established the height of the hill at various points. From these it was possible to sink shafts down to canal level and then work outwards from the shaft bottom and from the two ends. Headings were taken by making readings at the surface, dropping a plumb line down the shaft and then laying out the same headings at the bottom. Many of the problems were familiar from his work at Wet Earth. He soon found it necessary to install a water-powered pump to drain the workings, but as the works went deeper into the hill, he was forced to bring in a steam engine to provide more power. As the tunnellers moved deeper into the hill, ventilation also became a problem and a stove had to be lit to suck the fetid air out and allow fresh air to be drawn in. The clay that was excavated was burnt on site to make bricks to line the completed tunnel. But as the work progressed, so the difficulties mounted. There are no surviving detailed accounts of the period of construction, but in 1822, there was a complete survey, which described it as being in a sorry condition:

> In some places it is too narrow, in others crooked, and generally speaking the brickwork which forms the bottom, sides & top of the Tunnel is no more than 9 inches thick, it has throughout been made with bad mortar, so that in all the brickwork under water the mortar is as soft as clay.

In 1824 it was decided that rather than attempt to improve the old tunnel, a second tunnel would have to be built alongside it, to help speed the increasingly heavy traffic on the canal. The work was directed by Thomas Telford, who described what he found:

> The ground is of different and various kinds such as Rock, Sand, Coal Measures and other kinds of Earth – very tedious to encounter with. The Rock I find to be extremely hard, some of it in my opinion is much harder than ever any Tunnel has been driven in before – excepting that is executed by the side of it.

By the time Telford set to work, canal tunnelling had been going on for half a century; Brindley had no precedents to draw on. But much of the actual work was similar to working in mines. To break through the rock, the men had to drill holes by hand, pack them with black powder and blast their way through. It was not an exact science, so it is not surprising that the results were not altogether satisfactory either.

It is impossible to judge the quality of the work today, as the old tunnel has long since been closed and all the traffic uses Telford's replacement. It is, however, possible to compare the low narrow entrance of the original with the altogether more spacious Telford tunnel, which came complete with the luxury of a towpath. But the other tunnels along the canal, notably Preston Brook, show all too clearly that these pioneering tunnels were about as straight as a dog's hind leg.

Brindley's hectic life meant that it was impossible for him to keep up with the demands from all the different companies he worked for as engineer. There were grumblings from the Oxford Canal Committee, who recorded that 'Mr. Brindley hath in no Degree complied with the Orders of the Committee'. But when their engineer sent in his resignation, they retreated as fast as they could go. The Coventry Canal Company was the only one who actually had the temerity to sack the great man. But busy as he was, his life was not entirely devoted to work. One of his associates was the surveyor John Henshall, whose daughter was still at school. Brindley got into the habit of having a pocket of gingerbread for the schoolgirl when he visited, but as she grew older the relationship changed. She left school and in 1765 they were married; she was nineteen and he was forty-nine. It seems an unlikely pairing, but as any reader of Jane Austen will know, such matches were common at that time. There was, however, a rumour that she was not his first love, and that he may have had an illegitimate son, John Bennett, born in 1760. Brindley by now was beginning to suffer from bad health. He does not seem to have known that he was diabetic, a debilitating disease that eventually, combined with his huge work load, had seriously affected his health. In 1772, he was surveying the line of what was to become the Caldon Canal, when he got soaked to the skin and ended up in a damp bed at the inn. The resulting illness laid him low. He died that year at the age of fifty-six, leaving behind two daughters.

His old friend Josiah Wedgwood described his last days. He had visited Brindley every day for two weeks, but on 27 September he asked for a drink to wet his mouth, then said ''tis enough – I shall need no more', closed his eyes and never opened them again. The letter in which Wedgwood passed on the news contained a paragraph that could well stand as his written epitaph:

> What the public has lost can only be conceived by those who best know his Character and Talents, Talents to which this Age and Country are indebted for works that will be the most lasting Monuments to his Fame, and shew to future Ages how much good may be done by one single Genius, when happily employed upon works beneficial to Mankind.

There were other epitaphs of a different character. There were the engineers who learned their trade under his supervision: Hugh Henshall, Robert Whitworth, Thomas Dadford and Samuel Simcock, who carried on his work after his death. And, of course, there were the canals themselves, the start of a system that was transforming the country's transport network and feeding the burgeoning Industrial Revolution.

3

THE BROAD CANALS

THE NARROW canal system was developed on the pattern set by Brindley and the narrow boat carried its trade. But it was not the only system devised during the early years, nor was Brindley the only engineer. There were others, some of whom had very different ideas, and one at least who came from a very different background.

John Smeaton was born at Austhorpe Lodge near Leeds in 1724. His father was a lawyer and he received a conventional education, though his own interests were very far from usual. His daughter, Mary Dixon, in a brief memoir, described his early life from the age of six. 'All his playthings were models of machines, which destroyed the fish in the ponds, by raising water out of one into another.' After leaving Leeds Grammar School at the age of sixteen, John joined his father to study law. His heart was not really in it, and he was far more interested in spending his time with Henry Hindley, a brilliant clock and instrument maker, who he met in 1741. But he continued to follow the family's wishes and took himself off to London to continue his legal education. He stuck it out until 1744, when he returned to Yorkshire, abandoned the law and set about teaching himself the necessary mechanical skills to become a first-rate manufacturer of scientific instruments. It was a long process, but by 1748 he had established himself in London and was employing three assistants.

Smeaton's interest in instruments was only part of the story; he was no less fascinated by the ways in which they could be used. He was a practical rather than a theoretical scientist, and the experiments with which he made his name were closely related to the everyday world of industry. The principal power sources available at that time were water and wind, but no one had ever worked out the best way to use them, let alone produce any quantitative results. That was the task he set himself. He made his own equipment and devised his own experiments. It might be thought that as watermills had been in use for nearly 2,000 years, he was merely wasting his time finding out what everyone already knew. But what everyone thought they knew turned out to be wrong. Common sense suggested that a wheel set in a stream thundering along at a good pace must be getting more power than one that was simply turned by water dropping onto it from above. He found exactly the opposite to be the case. The most efficient wheel was overshot, in which the water fell into buckets on the rim of the wheel, allowing gravity to provide the motive force. The best of these had an efficiency of over 60 per cent, while the familiar undershot wheel, where the wheel was turned by the force of flowing water, only notched up a feeble 22 per cent. He presented his results to the Royal Society, received a medal for original research and in 1753 was elected as a Fellow. He was no longer merely an instrument maker, he was a member of the scientific elite – he was John Smeaton FRS.

John Smeaton, generally recognised as
Britain's first professional civil engineer.

Not surprisingly he was soon being
consulted on the construction of watermills and the instrument maker began
a transformation into an engineer. His
reputation grew and a catastrophe gave
him the opportunity to confirm his
reputation. There had been three light-
houses on the Eddystone rocks just off
Plymouth, all built of wood. The first,
designed by Henry Winstanley was rapidly replaced by a more elaborate version.
This was destroyed in a storm, taking
Winstanley with it. Its successor burned
down, set alight by its own lantern. The
next was entrusted to Smeaton. He was
determined that no such disaster would
happen to his work. It was to be built of
stone, with each large block keyed to the next by dovetail joints and the whole structure was
similarly keyed to the rocks on which it stood. That in itself was not enough and he began a
series of experiments to find a lime that would set under water, hydraulic lime. He found just
what he wanted by using limestone with a high clay content. It was a discovery that was to be
important to many other structures, some of which were to be part of the canal world he was
soon to join.

Smeaton was now in demand for all kinds of projects and was soon involved in navigable
waterways. He was put in charge of a plan to make the River Calder in Yorkshire navigable.
This was no easy matter. It was a fast-flowing river with a steep gradient. Taming it required
the construction of twenty-six locks with their weirs and 5 miles of artificial cutting in a total
length of just 24 miles. It was opened in 1764 as the Calder & Hebble Navigation. This was
engineering on a grand scale, with locks taking vessels over 57ft in length and 14ft beam, the
barges already in use on the earlier Aire & Calder; dimensions that were to prove crucial to a
scheme that we shall be looking at shortly. He also added many other engineering projects to
his repertoire, from bridge building to land drainage. He had become the complete engineer.
He not only did the necessary work, but also wrote his own elegant reports and illustrated
them with his own detailed, exquisite drawings and diagrams.

In 1764 Smeaton was called on to provide plans for a canal to cross Scotland from the Forth
to the Clyde, linking Edinburgh and Glasgow. It was not exactly a new idea. The first proposals
were made in the seventeenth century and a plan was actually produced in 1723, but was considered too problematical and far too expensive. The attractions of the scheme were obvious:
the only communication by water between the east and west coasts of Scotland was by the long
and hazardous sea route around the north coast. And by the 1760s Scotland was sharing in the
Industrial Revolution that was transforming the whole of Britain. The need for a canal was
now glaringly obvious.

Smeaton looked at two possible routes: one that began by heading north towards Loch
Lomond and then followed the low ground at the foot of the Campsie Fells and a second
that stayed further south and followed the Carron valley. His arguments in favour of the latter

The entrance lock and company offices on the Forth & Clyde at Bowling, where the canal joins the Clyde.

Barrels of whisky being loaded into a Clyde Puffer at Port Dundas, Forth & Clyde.

Pleasure steamers on the Forth & Clyde.

made it the obvious choice. His first report had to be made without any statistics on the likely annual tonnage to be carried by the canal, nor on the vessels that might be most likely to use the waterway. So he based his estimates on the 'ga' boats of the Clyde, 56ft long, 17ft 6in beam and 4ft draught. This odd name is probably an abbreviation of 'gabbert', the traditional sailing craft of the Clyde and the west coast – the name 'gabbert' was also adopted for canal lighters at a later date. He made it very clear that this was merely a suggestion on his part; it was up to the proprietors to tell him what their trade would be and what they thought would best serve the community. On that basis he made very careful calculations of the water that would need to be supplied to keep the canal open, laid out his line and estimated the costs. He explained his own role in words that are worth quoting at length. They show an innate modesty regarding his own standing and a very lucid view of what were – and equally importantly what were not – an engineer's responsibilities:

I do not act in consequence of any public authority or commission, to enquire, determine, and declare what is best for the public, or what in its consequences will assuredly promote its good. I should indeed be very sorry to have such great a weight laid upon my shoulders, which neither my studies nor inclinations have qualified me to undertake. I consider myself in no other light than as a private artist who works for hire for those who are pleased to employ me, and those whom I can conveniently and consistently serve. They who send for me to take my advice upon any scheme, I consider as my paymasters; from them I receive my propositions of what they are desirous of effecting; work with rule and compass, pen, ink, and paper, and figures, and give them my best advice thereupon. If the proposition be of a public nature, and such as involves the interest of others, I endeavour to deliver myself with all the plainness and perspicuity I am able, that those who may have an interest of a contrary kind may have an opportunity of declaring and defending themselves. I do not look upon the report of an engineer to be a law, at least not my own.

The proprietors now called for a second report to allow for larger vessels to use the canals. Smeaton produced new calculations and in 1768 the Act was passed. At this point, and before work had begun, the Company suddenly succumbed to an attack of the dithers. Was this really the best plan? Could the whole thing be done cheaper? Could they trust Smeaton? They decided to follow the example of everyone else involved with canal construction; they asked Brindley for his opinion. The great man duly appeared and wrote a separate report, which was then sent on to Smeaton. His reply was later published, and is worth reading at length as an example of a complete demolition job. Smeaton must have been furious at this implied lack of faith on the part of his employers, but satisfied himself with going through Brindley's proposals, point by point, and systematically refuting them. His pen dripped scorn, and it is tempting to quote at great length, but here are just a couple of examples. Brindley actually suggested that the whole scheme would be better if the wide canal was replaced by one of his new narrow canals. Smeaton pointed out that one of the major customers for the proposed waterway would be the Carron ironworks that specialised in casting cylinders for steam engines – big steam engines. 'I should be glad to see how he would stow a fire engine cylinder cast at Carron, of 6½ feet diameter, in one of his seven feet boats, so as to prevent it breaking the back of the boat, or overturning.' Brindley then proposed that work should start at the middle of the route, which Smeaton pointed out would be the most difficult to construct. He could not see any sense in starting inexperienced workmen at the most complex tasks, rather than have them start with the easy ground at the two ends. Smeaton considered why this perverse method of working was being proposed – was it merely because that was what Brindley usually did and what were the unspecified 'many accounts' that were vaguely mentioned?

> But pray, Mr. Brindley, is there no way to do a thing right but the way you do? I wish you had been a little more explicit on the many accounts; I think you only mention one, and that is to give more time to examine the two ends; but pray, Mr. Brindley, if you were in a hurry and the weather happened to be bad, so that you could not satisfy yourself concerning them, are the works to be immediately stopped when you blow the whistle, till you come again, and make a more mature examination?

Nothing more was heard of Brindley's proposed changes and one can only assume that there were a great many red faces among those who had commissioned his report. The canal was built to Smeaton's plan.

This was and is a majestic canal and the care in planning shows throughout. Smeaton had done his homework well, and unlike most British engineers of the time, he had looked overseas as well as at home for his models. He noted one major problem affecting trade on the Canal du Midi. It joins the River Garonne for its final run to the sea, but in summer the river frequently dries up, making it all but un-navigable. Smeaton made sure that his canal connected to the tidal reaches of both the Clyde and the River Carron that provides access to the Firth of Forth. There were a number of particular problems to be solved. For a start, the canal was intended for use by high-masted vessels, so all the bridges had to be swing bridges to let them pass. Although the canal was only 35 miles long, the summit was over 150ft above sea level, so there were twenty locks down to the Forth and nineteen down to the Clyde. All these had to be supplied by water and there were two reservoirs, the largest covering seventy acres. Not everything you can see today is directly attributable to Smeaton.

Work began under his direction, but the proprietors continued to interfere, seldom to the advantage of the works. Within three years over half the canal had been completed, but Smeaton finally lost patience and withdrew. The bureaucrats congratulated themselves on having a more economic alternative on hand. They decided they didn't really need a chief engineer at all. The plans had long since been completed and they entrusted the work to the main contractor, who could finish things off on his own. The result was disastrous. Work went on for a while, until in 1775 everything came to a halt. The canal had not been finished and worse still the coffers were empty. There was a gap of nine years before Robert Whitworth was given the task of completing the work. Ironically, he had learned his engineering under Brindley's tutelage, but he did a far better job for the canal company than his old master had done. One of the grandest structures, the Kelvin aqueduct, was actually the work of Whitworth and stands 65ft above the river.

Another canal was begun at much the same time and its chief engineer was a young man anxious to make his name in the world – James Watt. He was an instrument maker who worked closely with the University of Glasgow. He had begun to take an interest in steam engines, but had not yet arrived at a position where he could put his new ideas into practice. So, in the meantime, any other work was always welcome. He had been asked to help with the planning of the Forth & Clyde at an early stage, but his ideas were not followed through. He was, however, then asked to be chief engineer for a canal that would link the Monkland coalfields to Glasgow. It was a modest affair, but did exactly what it was built to do – provided cheap coal for the rapidly growing city. The canal was extended to join a branch of the Forth & Clyde at Port Dundas. There are few places in Britain where the importance of canals to the development of the country can be seen as clearly as here. The company offices are elegant and the waterfront is lined with imposing warehouses. By the time this development had occurred, however, James

Watt had moved on. He had formed a partnership with Matthew Boulton of Birmingham and together they became the most famous manufacturers of steam engines in the world. It was the steam engine that made his fortune and his reputation, but his early foray into canal engineering has also left its mark.

In later life, Smeaton was to work on other canals and navigations, not all of which got beyond the planning stage. One that did was the Birmingham & Fazeley that was to form a vital link between Birmingham and the canals leading to London. He also went to Ireland to survey the route for the Grand Canal. He was too busy to take over the role of chief engineer, but he recommended his young assistant for the job, William Jessop. We shall be meeting him again shortly. Looking at Smeaton's career as a whole perhaps his greatest achievement was to establish civil engineering as a profession rather than a trade, and to that end he started the Society of Engineers, generally known as the Smeatonians. It was as much a social institution as a professional body, with membership by invitation rather than qualification. Among those who were invited was a Yorkshire engineer, John Longbotham, who had started his working career under Smeaton's instruction. In some documents his name appears as Longbottom – perhaps a little gentrification had taken place somewhere along the way. His career was to have its triumphs but more than its share of disasters, and he was to experience both in his first major project, the Leeds & Liverpool Canal.

Smeaton did not have to concern himself with any other canals when planning the Forth & Clyde; it was never intended as part of an extended system. All he had to concern himself with were the broad waterways at either end. When it came time to think of building a canal across the Pennines, linking Liverpool on the Mersey to Leeds on the Aire & Calder Navigation, the same assumption was made. What no one anticipated was a waterways rerunning of the Wars of the Roses. Longbotham himself was an ardent supporter of the scheme from the first and persuaded others to take him seriously. From the start there were two committees – one in Yorkshire, the other in Lancashire – and after four years of argument they finally agreed what was needed and the Act was duly passed in 1766. For a time all was optimism; no one could have guessed that the canal would not be opened until 1816. The estimated cost was £259,777 – canal engineers never liked to round-off figures in case it looked as if they were not being accurate in their calculations. No one was really fooled, any more than pricing something at £9.99 convinces you it must be an absolute bargain at under £10. The committees might, however, never have started had they seen the final costs for the enterprise – roughly £1,200,000. Perhaps the most surprising thing about the whole project was the way in which it finished up – with locks at one end of the canal being a different size from those at the other. How could such a strange thing happen? That's where the old rivalries come in.

Longbotham conducted the first survey. The greatest obstacle he faced was the Pennine hills, which he did not exactly tackle head on, but rather tried to avoid altogether by taking a great, sweeping curve to the north through Skipton and Gargrave, some 16 miles to the north of Leeds. Inevitably, Brindley was called in to view the proposed route and, equally inevitably, he heartily approved of a plan that so closely followed his own favourite method of construction. Brindley was duly appointed chief engineer, and Longbotham given the job of Clerk of Works under him. But even Brindley realised that adding a route of nearly 130 miles through difficult country to an already overpacked schedule was a scheme too far. He resigned, and Longbotham moved up to take his place. Now the arguments really got serious.

The Yorkshire committee proposed an even more extravagant route that headed towards the Ribble and completely missed out all the developing textile towns of Lancashire, such as Blackburn and Burnley, with a vague promise that branch lines would be constructed at some future date. Neither side could agree so Brindley was called in once again, this time

Repair work to the Bingley five-lock staircase on the Leeds & Liverpool.

as arbitrator. The Lancashire men had called in an engineer of their own to devise a better line, but when Brindley inspected it he found that the levels were hopelessly inaccurate – at one point they were out by as much as 35ft. As Longbotham had proved himself perfectly competent by contrast, it was Longbotham's plan that was approved, and he was authorised to start at the Yorkshire end.

The all-important matter of locks had to be settled. The obvious factor to take into account was the size of the boats most likely to use the canal. In Yorkshire these were the barges already at work on the Aire & Calder, so Longbotham opted for broad locks to take vessels slightly over 14ft beam and 60ft long. The Lancashire men wanted the canal to connect with the Wigan coalfield and to be used by the Mersey flats, so they needed their locks 72ft long, and that's what they eventually set out to build all the way to Wigan. It now seems perfectly obvious that the greater length should have been allowed – not just so that Liverpool boats could travel the whole canal, but at that size the canal could also have taken narrow boats from the Midlands as and when the network spread.

Brindley may have approved the long route to the north, but it is doubtful if he would ever have devised anything as bold as Longbotham's plan for the Yorkshire end. The first part of the route lay up the valley of the Aire. The ascent began with the familiar spattering of locks, then as the valley steepened, locks were run together as pairs and in threes, a process that culminated in the last great leap up the hillside, the Bingley Five Rise. This is one of the most imposing engineering features on any British waterway, five wide, interconnected locks lifting the canal 60ft up the hillside. Everything about it is monumental, with lock chambers solidly constructed out of massive stone blocks. Once this point had been reached, life suddenly became a great deal easier. Longbotham could keep a level. He no longer followed the twists and turns of the Aire, but kept to a line above the river for 17 lock-free miles. As soon as this section of the canal was completed the company started making money and Longbotham began running a small fleet of passenger boats. Everything seemed to be going perfectly. Sixty miles of canal had been finished, boats were at work and everyone should have been happy. But the committee and Longbotham were about to part company. There is some doubt over the cause: some signs indicate that the money men were not happy with the rate of progress, others suggest there were discrepancies in Longbotham's accounts – though there was never any suggestion that he should either be prosecuted for fraud or asked to make good any mistakes. Whatever the cause, he was fired. If the quarrel really was about slow progress, then the committee must soon have regretted their action. It was a rerun of the Forth & Clyde story. Work slowed and in 1782 funds ran out and everything ground to a halt. Work was restarted under a new Act of 1790, and this time the Lancashire men had their say. The western end was rerouted and built with the longer locks.

Over the long years of construction, the Leeds & Liverpool acquired more notable engineering features. At Burnley it passes high above the town on a great embankment. Here, one can see how industrialists found canalside sites to be hugely beneficial, for the route at the end of the bank is lined by an almost continuous run of cotton mills. The flight of locks at Wigan is dauntingly imposing and one of the most miserable days I have ever spent on a boat was working up those twenty-one locks on an afternoon of unrelenting, freezing rain; no pub has ever looked more welcome than the one we went to that evening at the top of the flight. Perhaps one sometimes has to endure a bit of misery to make pleasures all the sweeter, and there is certainly a huge improvement once the Wigan flight has been conquered. I can think of few sections of canal that are more appealing than the Pennine section of the Leeds & Liverpool, as it swings through a rolling landscape chequered with fields bounded by mellow stone walls. Here it is easy to imagine that the canal was built purely to provide a wonderful experience for pleasure boaters, but there is no shortage of reminders that this

Loading stone into traditional short boats on the Leeds & Liverpool.

was first and foremost a commercial enterprise, serving industry and one can see what the canal meant to industrial development in Skipton. This is a traditional market town, spread out along a main street leading up to the castle. But move down to the canal and the whole scene changes. Textile mills follow the route and terraces of houses are crammed together to hold the newly recruited workforce.

There is a sad corollary to the story of this canal. In 1800 the Company received a letter from Longbotham. He had worked on several other schemes over the years, but now as he grew older he was all but destitute. He was not asking for charity, but simply a small stipend, for which he would offer his services as and when they were needed. The committee agreed that something might be done, but were in no great hurry to decide what that something might be. In the spring of 1801 they settled on the stipend. It was too late, he had died in the winter. All they could do was pay for his funeral. Some engineers may have made very comfortable livings indeed, but not all.

It would seem to be common sense that industrialists would be the first to welcome a new canal to their region, but it was not always so. In 1730 an Act was passed for making the River Stroudwater navigable from the Severn to Wallbridge, Stroud. At that time Stroud was at the very heart of the prosperous West of England cloth industry, just the sort of area where improved transport would be valued. The Act referred to the anonymous group of men who proposed building the canal as 'The Undertakers', a name that proved all too apt for the scheme was dead and buried before work ever got started. The assassins were the local mill owners. They had two objections. The first was that the navigation would take away water that was needed to power the mills. The second was that the waters of the local streams and rivers might become polluted. This was a serious matter for the dyers of the region, for Stroud was famous for its scarlet cloth – local mills were to provide the famous red coats of the British Army and continued to supply the Brigades of Guards until recent times. Those who wanted water transport were not so easily beaten.

In 1739 a new scheme was put forward by John Kemmett and three partners, which got round one of those objections, by proposing an early form of containerisation. This navigation would be lock free – so no water would be wasted. It was not going to be a continuous route, however. When it had to go up a level it came to a dead halt at a wharf, and the next section continued at the higher level beyond that. The wharf would have a pivoting crane that would lift the containers from one boat and load them onto the next on the upper level. Amazingly this cumbersome system was actually tried out and about 5 miles of river were 'improved'. Kemmett and his partners tried their best but were facing financial ruin when they called a halt.

Nothing more happened until 1774. Now the world of textiles was booming, transformed by the mechanisation of the processes for spinning yarn. Increasingly, spinning was being taken out of the cottages and moving into factories. These were still powered by water, but that was a problem that could be overcome by building a new canal, rather than simply straightening out the kinks in the river. This time it was a prominent clothier, William Dallaway, who led the promoters instead of heading the opposition. There were still complaints, but a new Act was passed authorising the construction of the Stroudwater Navigation in 1776 and work began immediately. Again, the dimensions were determined by the need to accommodate the typical craft of the region, the sailing barges known as Severn trows, so locks 70ft long by 15ft beam were constructed to take these vessels. Opened in 1779, it soon proved its value. Coal from the Forest of Dean was soon making its way to Stroud and the surrounding area. Mill owners began to acquire canalside sites for their wharfs and to construct new mills. Water transport began to have a more direct effect on the cloth industry itself. Samuel Skey had a chemical factory at Bewdley, the inland port on the Severn. Thanks to cheap transport he found a good

The castellated entrance to Sapperton tunnel.

A Severn trow being bow-hauled on the Thames & Severn.

market for his dyes in Stroud, which soon proved so successful and profitable that he ran his own fleet of barges. It all added up to an excellent argument in favour of canals, and people started asking the obvious questions. Should it be incorporated into the wider canal system? Should it be extended to reach the Thames?

The impetus for a Thames & Severn Canal came, however, not from the Stroud end but from the Thames, where barges were longer than the trows, but of narrower beam. This time there was no question of appointing Brindley, who had died in 1772, so the job of chief architect went to his old pupil, Robert Whitworth. He took the decision to opt for 71ft by 11ft for the locks, recognising that there would have to be an interchange point at the western end. This was to be Brimscombe Port, where the two types of barges could exchange cargoes. The one advantage of the schemes was that the canal could be used throughout its length by narrow boats – which was at least an improvement over the Leeds & Liverpool. There was to be a steady climb up the valley to a point near the village of Sapperton, where the hill proved just the sort of obstacle that Brindley had faced at Harecastle, and demanded the same solution. This was to prove altogether more challenging, for Whitworth was committed to a wide canal that demanded a wide tunnel, and it was to be by far the longest yet attempted. It was to be 15ft high, 15ft wide and 3,817 yards long and Whitworth wisely noted that the likely cost was 'an uncertain piece of Business', but he put in an estimate anyway with the usual dubious 'accuracy' of £36,575. If Whitworth showed himself to be unlike Brindley in being prepared to take bold decisions, he proved himself an apt pupil in other ways. Once he had completed the initial survey he was hardly ever seen again at the workings. The actual work was left to the resident engineer, Josiah Clowes, who deserves to be given most of the credit for building a canal through such a very difficult terrain.

Where most of the canals of the early period were commercial successes, the Thames & Severn faltered. It was badly affected by the opening of the Kennet & Avon, that offered a more convenient route for goods from Bristol, and the arrival of the railways made things worse. It might have weathered all these problems, had it not been for a continuous problem with water supply and leakages. The supply problem was partly solved in 1794 when a Boulton & Watt steam engine began work, pumping up water from a well at Thames Head. The canal might have been kept going, but in the nineteenth century it was taken over by the Great Western Railway, who had little interest in preserving a rival of any kind. And even when it was recovered from GWR control, the end was inevitable. By the end of the nineteenth century, although it was in a better state of repair, traffic was diminishing to near vanishing point. When E. Temple Thurston described his journeys round Britain's canals in his book *The Flower of Gloster* published in 1913, he met only one working boat on the whole canal. The lack of trade also meant he had to experience some hard work. Sapperton tunnel has no towpath, so boats were legged through. Until shortly before that time, there had been professional leggers to do the job, charging 5s for a loaded boat and 2s 6d for an empty one. After legging for an hour Thurston decided that 'A pound wouldn't satisfy me'.

After the passing of the Act for the Thames & Severn in 1780, only one new canal, the Birmingham & Fazeley, was authorised over the next eight years. It was not that canals were no longer viable but there was simply no money available. The War of American Independence brought a slump in trade that affected everyone. When confidence was finally restored, there was a steady stream of applications for new canals that soon swelled to a torrent. A new generation of engineers appeared, who had no shortage of precedents to build on, to which they added brand new ideas of their own.

4

THE NEW MEN

THE ACCELERATING pace of canal development can be seen in the numbers of Acts passed by Parliament that resulted in canals being built. The figures quoted do not include canals that were proposed but never got through Parliament and others that got their Acts but were destined never to be completed. In the years 1788 to 1790 there was just one Act each year, but there were six in 1791, six in 1792, nineteen the following year, nine in 1794 and after that the flow slowed to a trickle. This period was quite different from the earlier burst of canal building. Then the impetus had come mainly from industrialists, who were far more interested in obtaining a good transport route than in any dividends they might get from their shares – not that they were averse to making a profit, just that it was not their first priority. But the early canals did, in fact, make profits, sometimes very good ones and none did better than the Birmingham Canal. When they were first offered in 1767, shares could have been bought for £140 – fifteen years later they were on the market at £1,172. For many speculators buying a canal share looked like getting a lottery ticket with a win guaranteed. Any canal proposal was greeted with a flurry of activity as speculators rushed to get their names on the subscription list. They showed little interest in the nature of the proposal as one commentator noted:

> So unbounded have the speculations in canals been, that neither hills nor dales, rocks nor mountains, could stop their progress, and whether the country afforded water to supply them, or mines and minerals to feed them with the tonnage, or whether it was populous or otherwise, all amounted to nothing, for in the end, they were all to be Bridgewater canals.

The same account described how, as soon as news of a new canal project was announced, these speculators dashed to the scene, happy to sleep in barns and stables when the local inns were full. These men were only interested in one thing, getting in at the start, confident that they would soon be selling the shares on at a profit. So great was the demand that this period became known as the years of Canal Mania. It was a time when canals were promoted and built that were never likely to attract any traffic, nor pay a penny in dividend. The Melton Mowbray Canal is a typical example, wandering across the face of the countryside, reaching few towns and none with the sort of industries that were likely to provide profitable cargoes. There must have been a huge sigh of relief from the proprietors when the Midland Railway offered to buy it up, so that they could fill it in and build their tracks on top of it. It was the first time in half a century the shareholders had made a penny from the enterprise. Happily, not all canal ventures were so foolhardy, for this was also a time when canals of immense value were completed. A huge amount of construction work was thus crammed into a very short space of time, but the

William Jessop, chief engineer to some of Britain's most ambitious canal schemes.

first generation of engineers was no longer available to carry it through. It was time for new men to step forward. Leading the way and carrying the heaviest burden of work throughout this period was William Jessop.

Jessop's father was a foreman shipwright at Devonport Dockyard who was recruited by Smeaton to work with him on the Eddystone lighthouse. The latter described Josias Jessop as 'not a man of much invention', but praised his sound judgement and craftsmanship. William was born in 1745 and received a good education, excelling in French and the sciences, and showing even more skill in working with wood and metal. Whilst still a schoolboy he made a cello, which was said to be 'tolerably good'. At the age of fourteen he was taken on by Smeaton as an apprentice. This was a normal progression for a boy from his background, but it looked likely to come to an abrupt halt when his father died two years later. Boys had to pay for the privilege of being apprentices, and a man such as Smeaton would command a high premium for passing on his knowledge. Fortunately a family friend stepped in, and Smeaton was clearly pleased to keep the boy on. He described 'Bill Jessop' as 'a very good lad and I hope will so continue, and turn out so as to do me credit'. He was not to be disappointed.

Young William Jessop left Devonport for Yorkshire and progressed from being Smeaton's apprentice to being his assistant on all his many projects. His first big chance was the Grand Canal in Ireland, and soon he was being entrusted with other canal schemes. In 1777 two major events occurred in his life: he set up in business in his own right as an engineer and he married. He proved himself to be a conscientious and able engineer, hard working and modest. He seldom, if ever, looked for praise when things went well, but was always ready to accept responsibility when they went wrong. This trait appeared in the very first canal for which he was engineer at the start of the mania years, the Cromford Canal. This canal was something of an exception to the rule, in that it was very closely tied to industry and one particular industrialist, Sir Richard Arkwright. He was the inventor of the first successful cotton spinning machine that could be powered by a waterwheel and he had built his first mill at Cromford in Derbyshire. One of the reasons he had selected the site was its remoteness. He was well aware that the traditional spinners of Lancashire would oppose any invention that took away their livelihood or forced them to stop working in their own homes, with their family around them, in order to toil for long hours in a mill. The huge success of the Cromford mill made it inevitable that others would follow, and the Lancashire cotton manufacturers had no choice other than to join the rush into mechanisation. The new situation made Cromford's isolation seem rather less attractive: it needed connections with the rest of the world, hence the canal. It was to run from a terminus right outside Arkwright's mill to connect to the Erewash Navigation and through that to the River Trent.

The canal took Jessop into what was for him new territory: his previous experience had not really prepared him for building canals in such difficult country. He was in charge of constructing a tunnel almost 3,000 yards long and two major aqueducts across rivers. It was the latter that were to cause him grief; the aqueducts were disasters. Jessop was very cost conscious. The safest method of construction is to over-engineer everything, make doubly sure that the materials used were more than sufficient for the job. It must, however, have been very tempting for a young engineer on his first major assignment to prove to his employers that he could keep within budget – a feat very few managed. Sadly, he went too far. His first problem appeared when the aqueduct over the River Amber failed: the main arch with a span of 80ft gave way and Jessop bravely offered to pay for the repairs himself. The Company were happy to accept the offer. It was to cost him £650. Then the even grander aqueduct over the Derwent showed alarming cracks. Once again, Jessop accepted full responsibility. He explained that there was nothing wrong with his design, but that he had been let down by using cheap lime from the local quarries at Crich, which had failed to set. He wrote to the Committee that, 'I think it is common Justice that no one ought to suffer for the faults of another' and again offered to make the work good. It had not been an ideal start, but at least everyone now knew that when they dealt with William Jessop they dealt with an honest man.

After that somewhat difficult start, Jessop went on to become the engineer who was to carry the main burden of the mania years, taking on the role of chief engineer for some of the most important and most challenging canals of the era. Rather than give details of every canal on which he worked, I have selected just two to show how he coped with very different circumstances and different problems.

By far the most important canal for which he was responsible was the Grand Junction. By the 1790s, Birmingham had developed into one of the great industrial centres of Britain. It was linked by waterways to London, which as well as being the capital city was also the commercial centre of the country and its major port, but only by a very circuitous route. Boats leaving

Jessop's Derby Canal joins the Derwent, and the towpath had to be carried across the river on this low bridge.

Birmingham set off on their journey down the Birmingham & Fazeley into the Coventry and that in turn connected with the notoriously wriggling, wandering Oxford. Only then could boats enjoy the comparatively express route down the Thames. How difficult and problematical was this route? A good guide is to look up the distance from London to Birmingham on a modern road, which is roughly 120 miles: the unfortunate canal boatmen had to face a journey of 230 miles. Something clearly needed to be done. The older canals fought hard to keep their traffic, proposing a canal from London that would join the Oxford just north of that city at Thrupp. The rest of the journey would then have been as meandering as ever. A far bolder plan called for a route that would leave the Thames at Brentford, with an arm down to what was then the village of Paddington and connect with the Oxford at Braunston, 57 miles north of that canal's hoped for junction at Thrupp. This was the route that was approved in 1793. The Grand Junction almost exactly halved the distance between Braunston and the Thames at Brentford, and it was to offer other advantages as well.

The first stage of planning the route was shared between an older, experienced engineer, James Barnes and Jessop. By now the shortcomings of the original Brindley system were becoming all too obvious. Locks only allowed one boat at a time and narrow tunnels meant that there was no room to pass, so that boats might wait as long as half a day to get their turn. Add to that the fact that trade was steadily increasing and that the 25-ton load, that had seemed all but miraculous in 1766, was looking a good deal less so in the 1790s and the case for a more generous waterway was overwhelming. The two engineers took the decision to double the dimensions. It was hoped that this would be followed by the companies to the north doubling the size of their locks to allow river barges to pass along the whole route. That never happened, but at least narrow boats could share locks and pass in tunnels, and this was to prove vital in later years. They were planning for the future, which is never a bad idea.

A glance at the map is all that is needed to see that the canal is very different from its predecessors, taking an altogether more direct route. The first section, from the Thames, presented no great difficulties: there was an obvious line to take, following the valley of the River Brent. Jessop followed his old master's plan of making the junction with the tidal river, after which there is a comparatively steep climb of around 100ft, with most of the locks conveniently grouped closely together in the Hanwell flight. There is an odd – and definitely non-pc – name here, Asylum Lock. There is a clue in a bricked-up archway in a high wall beside the canal. This once allowed boats through to take coal to the Hanwell mental hospital. After that, things become easier and at Bull's Bridge, the Paddington Arm leads away on a lock-free journey of 13 miles. It was later to become still more important, when it was extended by the Regent's Canal to a basin on the Thames at Limehouse.

One of the features of the canal is the way that the bridges have been designed to a standard pattern, which is both economical and elegant, but there are occasional exceptions. Here at the junction, the bridge across the Paddington Arm has to carry the towpath of the main line, so it had to be made with a far gentler curve to make life easier for the horses to cross over it. This was a good, practical reason for not following the general pattern, but two other bridges along the way have a different story to tell. The canal had to pass close to the homes of two grandees, the Earls of Essex and Clarendon. They were prepared to tolerate the newcomer, but were able to demand that they were given bridges and structures that reflected the superiority of their surroundings. At the Essex estate of Cassiobury Park the bridge was built with a slightly incongruous Gothic arch and the same theme is repeated in the pointed windows of the nearby lock cottage. The Clarendons got something even grander at Grove Park, with a bridge that would have looked perfectly at home spanning an ornamental lake in the grounds. It is decorated with a cornice, balustrade and even sports a coat of arms.

The main line continues up the river valley, with a gentle rise through the locks, with no long flights, but all the time the canal is getting nearer to the long ridge of the Chiltern Hills. Here one finds the first dramatic evidence of the new, bold approach of the second generation of engineers. Where Brindley or Smeaton would have opted for a tunnel, Jessop chose to slice through the hills in a deep cutting at Tring. The advantages of a cutting over a tunnel came both during construction and afterwards when the canal was open for traffic. Unlike a tunnel, a cutting does not have to be lined with brick or stone and there is very little extra expense involved in making it wide enough to hold a towpath. It does, however, require the excavation of a vast amount of spoil, roughly five times as much as would have to be removed in building a tunnel. There were literally thousands of men at work on the canal at this time, and many were employed here digging away the earth with pickaxe and spade and blasting rock with black powder. It was not easy work; you do not have to dig down far in these hills before you reach the underlying chalk. The steep sides of the cutting made it all but impossible simply to wheel the spoil away, so barrow runs were used. These consisted of planks laid up the sides of the cutting on trestles. At the top of each run was a horse – when the barrow was full, it was hitched onto a rope and the horse at the top of the bank began walking away, drawing up the barrow. The workman's job was to keep the barrow – and himself – upright as he made his way up the slippery planks. It was dangerous work and the men soon learned that if they found themselves slipping, they had to make sure they fell to one side of the planks and pushed the barrow away to the other. It was bad enough falling down the slope without having a barrow load of rocks and clay landing on your head. Returning, they simply ran back down with the barrow behind them. There are no illustrations of the cutting during construction, but half a century later when Robert Stephenson came this way as engineer for the London & Birmingham Railway, he chose a line that closely followed that of the older canal, and he too slashed through the Chilterns at Tring. Fortunately there was an artist on hand to record the scene and it must have looked very similar when the canal was being constructed.

The Tring cutting marks the summit of the canal, almost 400ft above the start at the Thames. Each boat passing over it drained off approximately 56,000 gallons, which all had to be replaced. In this area water seeps easily though the chalk until it reaches the underlying clay, collecting to form an aquifer. This water could be reached by drilling boreholes, but a much easier solution was to find points where springs rose to the surface. The engineers found what they wanted near Wendover, so a navigable arm was constructed for 6 miles to tap this valuable source of water. Unfortunately, it was soon discovered that local mills were now running dry, so an alternative had to be found and the first of a number of reservoirs were constructed near the junction with the main line. This became one of the most important areas of the canal, essential for keeping the whole system running, and to help with that task a maintenance yard with extensive workshops was built, and the hamlet of Bulbourne grew up around it. The yard buildings are handsome, little changed and still very much in use, making wooden lock gates, just as they did two centuries ago.

Even greater obstacles lay up ahead, in the form of two tunnels at Braunston and Blisworth, 2,042 and 3,057 yards respectively. Braunston proved comparatively straightforward but Blisworth turned out to be a nightmarish experience. The tunnellers encountered all kind of problems, from quicksand to hard rock, and as if all that was not bad enough there was a calamitous collapse. Jessop was all for abandoning it altogether. He wanted to make a cutting at the top of the hill, approached on either side by locks, and kept in water by means of a steam pump. The idea was rejected, and Barnes' proposal for a tunnel on a different line was accepted instead. Work ground on, ever more slowly. To avoid a bottleneck, a plateway, a simple iron railway for horse-drawn trucks, was built over the hill to bypass the tunnel and provide the essential link between the two ends of the canal. The tunnel finally opened in 1805, over a decade after work had first started. Not surprisingly, Jessop's enthusiasm for tunnels, which had never been great, had by then vanished altogether. The result can be seen in the other great canal that he worked on at much the same time.

The Rochdale Canal served several basic needs. It was to provide a shorter route across the Pennines than the one offered by the meandering Leeds & Liverpool, which had still not been completed. It would also provide a direct link to the increasingly important town of Manchester, and the locks would be built long enough for use by both narrow boats and barges. This meant that it could form part of a coast to coast route, continued by the Bridgewater Canal in the west and the Calder & Hebble Navigation in the east. It has an interesting early history. The line of the canal was first surveyed by John Rennie, who we shall meet again later. Two proposals were put to Parliament in 1792 and 1793 and both were rejected. The arguments were the same as those that had sunk the earlier attempts to build the Stroudwater Canal – too much water was being taken from the streams that powered the local mills. At this point the company turned to the man of the moment, Jessop. He approved the basic route, but had severe doubts about the wisdom of having a long summit level that could only be achieved by having it run for much of the way in a tunnel. Jessop's first move was to get rid of the tunnel by having a shorter summit in a cutting and increasing the number of locks at either end. Most importantly, he looked over the surrounding countryside, tested the soil and declared that it would be possible to have the whole canal supplied through reservoirs. That would mean there was no longer any threat to the mill owners' water supplies and the main objection had been washed away.

This is the sort of statement that looks simple and obvious on paper, but takes on a very different aspect when you look at the ground that had to be crossed and the work involved. Samuel Smiles, in his *Lives of the Engineers* of 1862, compared the work on the canal with George Stephenson's Manchester & Leeds Railway and declared them comparable achievements. 'Whoever examines the works at this day – even after all that has been accomplished in canal and railway engineering – will admit that the mark of a master's hand is unmistakably stamped upon them.' Having walked the whole length of the canal and returned again by train I can only agree – it is just a shame that Smiles put Rennie down as the chief engineer, and never mentioned Jessop. It is certainly fair to give Rennie the credit for selecting a good route over difficult country, but it was Jessop's plan that was accepted and it was he who supervised the construction. So what are the features of his canal that make it so striking?

The canal is only 33 miles long, but in that comparatively short distance there are eighty-three locks, not narrow locks, but major works able to take vessels up to 74ft long and 14ft 2in beam. To keep water loss to a minimum, Jessop tried to keep the rise and fall as even as possible, at an average of 10ft per lock. The greater climb was from Manchester to the summit, which with a rise of slightly over 435ft in forty-five locks shows just how close he came to achieving his plans. It is not difficult to admire the workmanship of this remarkable canal, not least the attention to detail. Although the main supply was to come from reservoirs, there were a number of small streams that trickled down the Pennine hillsides along the way. These were channelled down into the canal, but only after passing through a stone trough that allowed any dirt and grit to settle out first. The engineer also faced a problem that had not been encountered on an English canal before – the need to cope with a road that had to cross the canal at an angle. It is very easy to build a bridge at right angles to the canal, and the simplest solution is to put a kink in the road at either side of the conventional bridge. Jessop, however, opted for a far more sophisticated solution – the skew bridge. In this the stone courses follow a spiral. It is one of those things that is easily overlooked, but once you have seen one it seems almost miraculous that anyone could produce something at once so intricate and so visually satisfying.

There are places where the struggle against a hostile terrain is all too evident. At the summit itself the canal passes close to the hill tops, where the dark Pennine rock breaks through the thin, moorland soil. Even more dramatic is the section between the two mill towns of Todmorden and Hebden Bridge. Here the canal is squeezed into a narrow passage between the hills. The route hugs the steep hillside and one can still see where the rocks were shattered by gunpowder to make way for it. But the most impressive engineering features of all are not even seen when travelling the canal itself.

Jessop's original plan called for two reservoirs, one at Hollingsworth and the other at Blackstone Edge, though he later found it necessary to build a third in the Chelburn Valley. We don't have a record of Jessop's surveys to find the sites for his reservoirs, but it is interesting to speculate if, when he was looking at the area round Blackstone Edge, he took time off to look at the work of a much earlier generation of civil engineers, who had forced a route across these bleak hills. Less than a mile to the south of his reservoir is a substantial and well preserved length of Roman road. His own task in building the reservoirs was every bit as daunting as that facing Britain's first road builders. Water exerts great pressure and he had to contain it behind substantial earth banks. The main bank at Hollingsworth is 10ft wide at the top and then slopes down on a 1:2 ratio. It was essential to keep it watertight, so at the core is a 9ft-wide core of puddled clay. The need for such elaborate measures was demonstrated a few years later on the Huddersfield Narrow Canal. This was the third cross-Pennine route and also required extensive reservoirs to keep it supplied. On the night of 29 November 1810, the dam wall of the Swellands reservoir gave way. Fortunately it had not yet been completely filled, but even so the water surged out with such force that a 15-ton boulder was washed away and ended up 2 miles down the valley. The waters flooded houses and mills in the Colne Valley and five people lost their lives. It became known as 'the night of the black flood'. No such disaster overtook Jessop's reservoirs, which with time blended into the landscape and even became popular beauty spots. Hollingsworth reservoir is now Hollingsworth Lake and it is probable that few of those who go boating on its waters or picnicking on its banks ever think of it as a triumph of eighteenth-century engineering.

One measure of Jessop's success can be gauged from comparing the fate of the Rochdale with the other two canals across the Pennines. The Leeds & Liverpool, begun in 1770, was finally opened in 1816. The Huddersfield Narrow was in many ways similar to the Rochdale in that it took a route through the heart of the Pennines. This time, however, it was decided to shorten the route to just over 19 miles from Huddersfield to Ashton-under-Lyne. It has the highest summit of any British canal and a total of seventy-four locks, but these were narrow boat locks, so from that point of view the construction problems should have been considerably less than those of the Rochdale. The difference lay in the central passage through the hills. Standedge Fell rises right across the line of the canal, reaching a height of almost 1,500ft. The decision was to go through it by tunnel, and it was to be the longest tunnel ever built on any British canal, a staggering 3 miles 135 yards long. Not surprisingly the engineers were keen to do no more excavation than was absolutely essential: it had no towpath so boats had to be legged through. The advantage of the short route was supposed to outweigh the disadvantages, and it could well have competed with the Rochdale for trade if it had been completed on time. The two canals were both begun in 1794, but the Huddersfield was not open until 1811, by which time the Rochdale had already been in business for seven years. Jessop's views on the subject of tunnels had been more than justified.

These were by no means the only canals Jessop was working on at the time and it was impossible to give his full attention to all of them. The Grand Junction was certainly given a very high priority and on other schemes he had to rely on the services of a skilful and reliable resident engineer to take on the full-time responsibility of overseeing the works on the spot. One canal which was situated a little way down the pecking order was the Ellesmere.

Chester had once been an important port being served by the River Dee, but by the eighteenth century traffic had sorely diminished, and complaints that trade was ruined were appearing in publications as early as 1677. The city fathers did nothing very much about it for the next hundred years. One way to increase trade was by new canal connections, and the Act for the Chester Canal was duly approved in 1777. Even then it was a canal to nowhere. It was hoped to link in with the rest of the newly developing system, but in the event it wandered across the Cheshire plain and came to a halt at the old salt town of Nantwich. That was the situation when Jessop was called in to consider how the canal could be extended, preferably to join the Dee to both the Severn and the Mersey and to take in the valuable iron works and coal mines of North Wales. The engineer decided that this was far too ambitious and settled for a canal that would join the Chester near Nantwich and link in with the industrial region near Llangollen. The only town of any note it passed through en route was the modest market town of Ellesmere which, if it did nothing else, gave the infant canal a name. It duly received its Act of Parliament in 1793. Jessop had surveyed the route, laid his plans – though leaving at least one major problem to be solved later – and looked for an engineer to join him as his right-hand man. Three men had worked with him and one of them, William Turner, was all but certain he had the job in his grasp. It was not to be. The heart of the new enterprise lay in Shropshire and the local bigwigs had recently acquired a new County Surveyor who was making a very good impression. He had no experience whatsoever in canal work, but he was appointed over Jessop's head. His name was Thomas Telford.

Telford seems the perfect fit to the type idealised by Samuel Smiles, the man of humble origins who raised himself to fame and fortune by his own efforts. To a very large extent the picture is accurate, but at various key points in his life he had advanced through influence and patronage – and this was one of those points, though he more than justified the faith of his patrons.

He was born in 1757 in a cottage in a peaceful valley in the Scottish Lowlands, near Eskdalemuir. His father was a shepherd who died just three months after Thomas was born. This was a double disaster for a young family. Their home was owned by the farmer and was now needed for the new shepherd, so mother and baby had to move into a single room in another nearby cottage. His mother, Janet, seems to have been a proud woman and it must have been painful for her to accept charity, but she had no choice and had to depend on her brother to keep them. The boy went to the local village school for a rudimentary education, but it was inevitable that as soon as he was old enough he would have to begin earning his living. He was apprenticed to a stonemason in Lochmaben. It did not last. He was cruelly treated and once again Janet had to appeal to her mother's family, the Jacksons, for help. As steward to an important local family, the Johnstones, Thomas Jackson had influence, and he found young Thomas a place with another mason in Langholm. It was to prove a doubly advantageous move. His new master was conscientious and a good teacher, and he found himself in an area where big improvements were under way. The local landowner, the Duke of Buccleuch, had put in hand all kinds of works. He had begun replacing cottages that had been little better than mud walls topped with heather thatch with new ones of stone with slate roofs, and he was improving the local transport system by rebuilding the roads and erecting bridges. There was ample opportunity to learn a whole range of skills. At some time around 1778, Telford was involved in the construction of the very substantial bridge across the Esk at Langholm. He had ended his apprenticeship and was entitled to put his own mason's mark in the stonework – it is still there.

These days were looked back on with great fondness by Telford. It was not all work. A local lady, Elizabeth Pasley, was very keen on improving the literacy of the local people and encouraged anyone who was interested to make use of her own library. Telford proved an enthusiastic pupil and he discovered in himself an immediate response to poetry. Inevitably he would later extol the virtue of Scotland's poetic hero, Robert Burns, but the book that first caught his imagination was Milton's great epic *Paradise Lost*, not a book one would have thought an automatic first choice for many teenage boys. In time, he was to write poetry himself and to be the friend of such distinguished writers as William Cowper and Robert Southey. He never forgot his beginnings as a simple mason nor the kindness that he found in Langholm. Many years later, when he was a famous engineer, he returned to the area, taking with him the tool kit he had kept through the years, and there he practised his old skills, carving two headstones. One was to be the stone that the family had never been able to afford to mark his father's grave, the other was for the Pasley family. You can still see the stones, which also provide good evidence of his skills, for the lettering is crisp and neatly formed.

He was a young man of talent, with ambitions that were never going to be realised in rural Eskdale. In 1780 he set off on foot to look for work in Edinburgh, which was then in the middle of a building boom. There he had the chance to learn not just by copying the work of others, but by reading and studying. He was, he felt, ready to go even further. In 1782 he set off for London. Here he rapidly found a job on the most prestigious building project in the city, Somerset House. But it was not London that was to help him advance his career, but his old Scottish connections. He was invited to work on improvements for a mansion in Scotland for Sir James Johnstone, who had married into the Pulteney family of Bath. This was the family after whom the well-known Pulteney Bridge over the Avon was named. Sir James was a man of considerable influence, who seems to have been delighted to find a man who was employed as a mason, who could understand architectural drawings, was highly literate and an amusing conversationalist. It was to prove an important connection.

The next major step in his career came when he was invited to work on the reconstruction of the Portsmouth Dockyard, but this time as an overseer not just a workman. He was there for two years, but he kept the Pulteney connection. When William Pulteney was elected as MP for Shrewsbury he decided that the run down, ramshackle castle was just the sort of home the local MP should have. He called for Telford and offered him the job of renovation. In 1786 he appeared at the castle as temporary master and chief architect. This was one of the rare periods of his life when work was not all consuming. He found time to write verse and read widely. He moved among the social elite of Shrewsbury. One of these, Katherine Plymley, wrote warmly of him as 'a most intelligent & enlightened man, his knowledge is general, his look full of intelligence & vivacity. He is eminently chearful & the broad Scotch accent that he retains rather becomes him.' He went to the theatre and greatly enjoyed a comedy, but concerts left him totally unmoved – 'I might as well have stayed at home.' Most importantly, he was now known to all the local people of influence and he made such a good impression that when the post of County Surveyor for Shropshire became vacant, he was the obvious first choice.

This was a job full of opportunities. He had always had an ambition to work as an architect, designing buildings, not just putting the stones in place. Now he had the chance. He worked on the new gaol and designed two churches, one at Bridgnorth, the other at Madeley. Both still exist and are perfectly competent examples of the current style of rather frigid classicism. He also had the opportunity to design a bridge across the Severn at Montford and even got involved with the excavation of the Roman remains at Wroxeter. It was a busy and satisfying life, and he could well have continued in it if the Ellesmere Canal had not needed an engineer. It was the Shropshire proprietors of the canal who put his name forward, and as they were paying the bills, Jessop had to accept this young man of whom he knew nothing whatsoever. Telford was very excited about his involvement in what he described as 'the greatest work in

An alternative to locks: the Trench incline on the Shrewsbury Canal – tub boats are floated onto wheeled cradles, which run on the rails (above). The steam winding house is at the top of the incline.

the kingdom'. At the same time he still made clear that his contract allowed him to return to take charge of any important works in Shropshire that might come up. He was clearly not yet certain which way his career was to go. It would soon be obvious. He would not, after all, be a second-rate architect; he would be a first-class engineer.

The Ellesmere Canal is a magnificent example of how Jessop was able to use a whole range of engineering ideas to meet the demands of a difficult terrain. Today the canal is known as the Llangollen, but that town was only included in the scheme because of the need for water supply. A weir was constructed across the River Dee, known rather romantically now as the Horseshoe Falls, and a feeder was constructed through Llangollen itself to join the main canal. As there were limestone quarries near the route, which could supply a valuable cargo, it was decided to make this feeder canal navigable – if only just. It is narrow and twisting, hugging the hillside, and though it must have been inconvenient for working boats, it has become one of the most beautiful and delightful sections of any canal for pleasure boaters. Unusually, this also marks the summit of the canal, from here it is downhill all the way, physically, but certainly not metaphorically. This is a canal packed with interesting features.

Starting in the east, the main line leaves the Chester Canal at Hurlestone Junction and immediately climbs up from the plain through a short flight of locks before continuing in a very conventional manner, heading south-west and gently rising through a spattering of locks. From Wrenbury the route turns more westerly and as the countryside gets hillier so the line becomes a little more wayward, but maintains its steady progress, until it reaches Grindley Brook, where it charges up the hillside via a three-lock staircase. The lock cottage is of interest, for it shows Telford exercising his architectural skills. In general lock cottages were kept plain, simple and functional, but this is altogether grander with a wide bow front and handsome verandah. Having reached a new level, the canal makes an almost right-angled bend and shortly afterwards exercises an almost complete U-turn. This very untypical layout has a rational explanation. This is not a thickly populated part of the world, and one of the few towns of any size is Whitchurch. It made sense to make a wharf near this important centre and it was considered worth the detour; in time the company went a step further and added a short branch down into the town centre.

The engineers now faced a more daunting challenge. Whixall Moss was a large area of boggy peat, so the first task was to cut drainage channels to create a reasonably firm foundation. Then an embankment was built up on the dried out land and the canal cut in the top of it. Having got past that obstacle the canal now arrived at The Meres, a sort of miniature Lake District, through which the canal threaded its way. It eventually reached Ellesmere itself, where the company had its headquarters building with a short arm down to an extensive town wharf. A little way beyond that is Frankton Junction. Most people nowadays regard the bit of the canal that carries straight on in the same direction to be the main line, and are then slightly mystified to find bridge numbers starting over again at Number 1. But the actual main line is the route turning south to connect with another canal authorised at much the same time, the Montgomery. The rest of the route towards Llangollen is no more than a branch line, hence the numbering.

Travelling on this part of the route today, is a delight, but when the canal was new this was an important coal mining region and the waterway was heading toward a busy industrial area of ironworks that would provide it with most of its paying customers. Before they could be reached, however, there were a number of major obstacles to overcome, of which the most demanding were the crossings of the two river valleys of the Ceiriog and the Dee. Long after work had started on other sections of the canal the debate on how this was to be achieved was still going forward. The answer was to be born out of accident and tragedy.

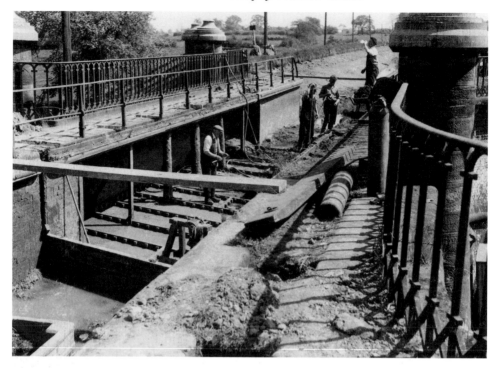

Repair work on the Shropshire Union Canal reveals the construction of an iron-trough aqueduct.

The Shrewsbury Canal was not exactly a major undertaking at a mere 18 miles long and the only engineering work of note was a crossing of the River Tern at Longdon. The river itself was not navigable, so it presented no real challenge to the engineer Josiah Clowes and work began on a conventional aqueduct. Before it was completed Clowes died and a sudden storm washed away the work that he had begun. Telford was seconded to the Shrewsbury to complete the canal and the aqueduct, and the obvious solution for the latter was simply to start again. But this canal was rather special. It was to be connected to the Ketley Canal, built to serve the iron works of William Reynolds, and it was Reynolds himself who had acted as engineer for the latter. This was a short canal with a severe gradient and Reynolds realised that the conventional solution of building lots of locks was impractical. Instead he built an inclined plane. The boats on the canal were very much what their name suggests – tub boats, little more than floating boxes. When a boat reached the incline it was floated onto a wheeled carriage that could be raised and lowered along a railed track laid on the hillside, and was then floated off again at the other end. Reynolds was not a man afraid of innovation. He was also a man keen to find new uses for the iron on which his livelihood depended. He and Telford put their heads together and came up with a new idea. Why not make an iron aqueduct? The structure was to consist of a cast-iron trough, made up of twenty-six sections bolted together, with a towpath slung on the outside. It was to be supported by three triangular iron piers on masonry bases. The canal has gone, but the aqueduct has been preserved. Visiting the site, one is struck by the contrast between the massive stone abutments of Clowes' original conventional structure and the seemingly fragile iron work. There was a lesson here that was not lost on Telford and Jessop back at the Ellesmere.

Of the two aqueducts that would be needed on the Ellesmere, by far the largest would have to be the one over the Dee. The canal as planned was to arrive at the Dee valley at a

Pontcysyllte, carrying the Llangollen Canal across the River Dee.

point 125ft above the river and the next point at the same height across the valley was over a 1,000ft away. To build any aqueduct in stone to span this gap was an immense challenge as the piers would have to support not only the trough and its water but also the essential lining of heavy clay. The best idea anyone had come up with was to have locks going down each side of the valley to a point where it would be more feasible to span it. The aqueduct would have acted as a sump, so steam engines would have been needed to pump water back up to the tops of the locks at both ends. But use an iron trough and everything looks very different. The iron trough itself would be far lighter than one built in masonry and it would have no need for a lining. The problem had a solution. The whole valley could be spanned by the one aqueduct that could be carried on nineteen arches, each with a 45ft span. Because the trough itself was so light the piers did not have to be completely solid; the upper sections were built hollow with cross bracing. The result was Pontcysyllte, arguably the single most spectacular engineering feature of the whole canal age.

There still remained the Ceiriog crossing to worry about. The Chirk aqueduct across the river appears completely conventional, but that too is actually an iron trough, given a thin outer coating of masonry, which is largely cosmetic. The approach to Chirk is via a massive embankment, and in the short distance between here and the Dee valley, the canal plunges into deep cuttings and a short tunnel. This really is impressive engineering. It would have seemed a triumph on any other canal, but it has been dwarfed by Pontcysyllte.

There has been a good deal of controversy over the years as to who should be given the credit for the construction of Pontcysyllte. Contemporaries lavished praise on Telford and when he had his portrait painted in later years, Pontcysyllte formed the background. But it always has to be remembered that Jessop was still the chief engineer and the decisions on what should be built and how ultimately rested with him. It was a subject that I argued about for many years with the doyen of canal historians, the late Charles Hadfield. He was a great admirer of Jessop and I tended to favour Telford. In the end we agreed that Jessop had been given less credit than he deserved, which meant that Telford had been overpraised and the truth probably was that the credit should be shared. It was a very British compromise. But on whichever door the laurels should be hung, there is no doubt that Pontcysyllte made Telford's reputation and instantly raised him into the top rank of British engineers. His canal engineering career could scarcely have had a better start.

Telford was to work on a number of canal schemes over the years, but by far the most ambitious took him back to his native Scotland and was part of an overall plan to raise the living standards of the Highlands. It was a scheme that he embraced with enthusiasm. The Highlanders

Thomas Telford being driven to inspect the works on the Caledonian Canal. Most of the manual work was carried out by Highlanders, like the two to the left of the picture.

had suffered greatly in the aftermath of the Jacobite rebellions and it was widely recognised that if the region was ever to develop economically it would need a proper transport infra-structure. Telford would be involved in the construction of new roads and ports, but the biggest single proposal was for a ship canal across the middle of the country. The need for the canal was explained by Sir Edward Parry, who gave evidence in its favour before Parliament. He cited the example of two ships leaving Newcastle on the same day: the vessel heading for Bombay arrived before the one making the trip round the north of Scotland reached Liverpool. There had always been an obvious route, following a natural fault line, known as the Great Glen, link-ing Fort William to Inverness. Loch Linnhe cut a deep bite into the west coast and an equally large chunk was carved out by the Moray Firth on the east. In between were three lochs that could be made navigable, Lochy, Oich and Ness. Link those by artificial canals and the job was done. It was not a new idea. James Watt had actually carried out a survey in 1773 and had esti-mated that a 10ft deep canal could be built for £164,000, but nothing was done. The impetus for change came from the Government. Poverty was driving more and more people out of the Highlands, many of whom were heading for a new life in Canada. In a report of 1803 it was reported that 3,000 had left in the previous year and three times that number were planning to join them, but that they would rather stay, if only there was work. The canal suddenly had a double value – the shipping route was needed and it would supply work for the Highlanders. It was not unlike an early version of Roosevelt's New Deal. It was enough to get the Act passed in 1803 and work on the Caledonian Canal could begin.

This was, by the standards of the age, engineering on an unprecedented scale. The canal had to be big enough to take merchant ships and naval frigates, so allowance was made for vessels up to 150ft long, 35ft beam and 13ft 6in draught. The locks had to be huge and the channel itself had sloping sides, with a 50ft width at the bottom and over 100ft at the surface. Watt's estimate was discarded and a new cost suggested of just under £500,000. It was to end up costing twice as much. Because this was a Government-backed project, we actually have a lot more informa-tion than usual about the progress of the works through the published annual reports, and it is also one of the few canals where a snapshot of work in progress was provided by a well-known literary figure. Telford was friendly with Robert Southey and the poet joined him for a tour of the works. He compared the canal to the pyramids of Egypt – to the latter's disadvantage:

In them we would perceive only a vain attempt to vie with greater things. But here we see the powers of nature brought to act upon a great scale, in subservience to the purposes of man: one river created, another (and that a huge mountain-stream) shouldered out of its place, and art and order assuming a character of sublimity.

He wrote these words when visiting the western end of the canal. The first challenge was to link the sea loch to Loch Lochy. It needed no more than 8 miles of artificial cutting, but demanded a vast amount of work. First the canal had to be thrust out into deep water, and a sea lock constructed so that it could be easily accessed apart from a period of two hours either side of low tide. The greatest obstacle was the need to raise the canal over 6oft in a very short distance, so the bold decision was taken to do it via eight interconnecting locks, which became known as Neptune's Staircase. They are not just dramatic in themselves but have a romantic setting, overlooked by the massive hunched shoulder of Britain's highest mountain, Ben Nevis. The rest of the way to the loch may look straightforward, but in reality it involved rerouting the river into a new channel – Southey's mountain-stream – and the construction of an aqueduct over the River Loy.

Even the lochs presented their own challenge. It was necessary to dredge out a navigation channel and steam dredgers were called in to help with the work. The next short section of canal proved to be among the most difficult of all. It was decided to pierce the high ground between Lochs Lochy and Oich by a deep cutting at Laggan. Given the huge amount of excavation needed, the work involved here was unlike anything attempted before. Instead of the barrow runs of the narrow canals, a far more sophisticated system was used. Southey described the scene:

A steamer descending Neptune's Staircase, Caledonian Canal.

The earth is removed by horses walking along the bench of the Canal, and drawing the laden cartlets up one inclined plane, while the emptied ones, which are connected with them by a chain passing over pullies, are let down another… The hour of rest for men and horses is announced by blowing a horn; and so well have the horses learnt to measure time by their own experience and sense of fatigue, that if the signal be delayed five minutes, they stop of their own accord, without it.

The 'cartlets' were specially designed for the job. They were built on a triangular frame, the wheels attached to the struts parallel to the slope, so that the platform remained horizontal on its journey up and down the incline. The work proved so difficult that a compromise was needed; the width of the bottom of the canal was reduced to 30ft. Beyond that there was a steep drop down to Loch Ness via the five-lock staircase at Fort Augustus. At least there were no worries about deep, long Loch Ness. After that there was just 5 miles to go to the sea, but the problems were not yet at an end. The coastal waters are shallow, so if ships could not get in to reach the canal, then the canal would have to be pushed out to reach the ships. An embankment had to be built for about 400 yards from the high-tide mark. The top of the bank was then loaded with stones and left for six months to settle. After that the canal could be cut into it and the sea lock constructed. The whole canal was finally opened in 1822. It would be good to report that it was a huge success. It was not. No one had foreseen that the age of comparatively small, wooden sailing ships was to give way to large, iron-hulled steamers that did not have to wait for a favourable wind to take them round the coast. It soon became all too evident that cost-cutting had resulted in some shoddy workmanship and by the middle of the nineteenth century a lot of rebuilding became necessary. Even so, it remains a considerable engineering achievement.

Telford's work in England saw him in the more familiar territory of narrow canals. The Birmingham & Liverpool Junction Canal was specifically designed to provide a short, fast route between Birmingham and the Mersey. Until then boats had to follow a tortuous route along the old Brindley canals. The new route was made as direct and straight as engineering would allow. Here one can see the technique of 'cut and fill' taken to extremes, with deep, gorge like cuttings, followed by high embankments. It epitomises the new thinking of the late period of canal construction. But to see just how much the new technology differs from the old, the ideal place to go is the main line of the Birmingham Canal itself. Telford designed a brand new canal that sliced straight through the wavering lines of Brindley's original. The result was a wide waterway, running straight and true and marked by deep cuttings. The old canal, however, still served a number of works and factories that were valued customers, so it was kept as a series of loops from the main line. A typical example is the Soho Loop, which served the Boulton & Watt factory. The improvements ensured that Birmingham retained its importance at the heart of the whole narrow canal network.

The other great Scots engineer of Telford's generation was John Rennie, though he came from a very different background. He was born in 1761 into a comparatively well-off family: his father was a farmer and brewery owner. The boy, however, received a somewhat unconventional education. It started off quite normally at the local school, but he showed such an enthusiasm and aptitude for all things mechanical that, at the age of twelve, he went to work for a local millwright, Andrew Meikle. He was no ordinary millwright, but the inventor of the first successful thrashing machine and Rennie spent two years with him before returning to school, satisfied that he now had a good practical grounding in mechanics. He proceeded to hop between formal education and the practical life, setting up in business as a millwright and

A typical deep cutting on the Shropshire Union.

Old canals were improved. Telford built a new straight main line for the Birmingham Canal, crossed by the splendid Galton Bridge. A typical narrow boat can be seen on the left, and a day boat on the right.

then attending Edinburgh University. His son, himself an eminent engineer, Sir John Rennie, later described his father's views on education:

> My father wisely determined that I should go through all the gradations, both practical and theoretical, which could not be done if I went to the University as the practical parts, which he considered most important, must be abandoned: for he said, after a young man has been three or four years at the University of Oxford or Cambridge, he cannot, without much difficulty, turn himself to the practical part of civil engineering.

But if John Rennie got nothing else out of his time at Edinburgh, he did make some useful connections, making friends with the eminent chemist Professor Joseph Black and John Robison, both of whom had known and helped James Watt in his career. Armed with an introduction from Robison he went down to Birmingham to the Boulton & Watt factory, where he was at once offered a job. He was sent to London to oversee the construction of the Albion Mills in London. Building a flour mill may not seem a very exciting occupation, but this was to be a mill unlike any other. At that time the biggest mill in London was worked with just four pairs of stones. This was to have thirty pairs, and was to be powered by three steam engines, not by waterwheel. It created a sensation and crowds came to see the first engine at work in 1786. The second engine was up and running by 1788, but in 1791 the building caught fire and was totally destroyed. Nevertheless, Rennie had played an important part in one of the key technological advances of the Industrial Revolution: the application of steam power to working machinery other than just pumps. He was clearly a man to watch, and it was not long before other major commissions were coming his way. It is still a little surprising to find him suddenly turning from mechanical to civil engineering, when he was appointed chief engineer for the Kennet & Avon Canal.

The canal was built to connect two existing river navigations: the Kennet joining Newbury to the Thames at Reading and the Avon from Bristol to Bath. To ensure continuity it had to be built with broad locks and it was no simple introduction for a canal-building novice. The most dramatic feature is the flight of twenty-nine locks lifting the canal up the hillside at Devizes. He was well aware of the need to conserve water, so the locks have side ponds terraced into the hill to act as a series of mini-reservoirs. But he was still faced with major problems of water supply and in solving them he showed what he had learned in his earlier working life as a millwright. After a steep climb up from the Avon at Bath, there is a long summit that needed to be kept supplied. Rennie's solution was to acquire an old mill site on the river and build a new pumping station. A pair of breastshot waterwheels, each 15ft 6in diameter and 11ft 6in wide, were installed to work two beam pumps, powerful enough to lift the river water 50ft up the hillside at a rate of 100,000 gallons an hour. They were later replaced for everyday use by an electric pump, but when that failed in 1989 the old system proved perfectly able to cope with the emergency. The next requirement was to supply the summit level, and here Rennie opted for a different solution: he turned to his old employers, Boulton & Watt, for a steam engine to work a pump at Crofton. A second one was added later. The older of the pair, that was supplied in 1812, is now the oldest steam engine in the world that is still on its original site and still able to do the work for which it was built. They demonstrate just why steam replaced water as a power source: these two working together can shift a million gallons an hour. I have been visiting Crofton for very many years and I never tire of watching these nodding, hissing giants at work.

There is another aspect of Rennie's work that is shown on this canal – the elegance of his designs. In part of the route he had no option. On its way out of Bath the canal had to pass by fashionable Sydney Gardens and to the ladies and gentlemen strolling its gravel paths the idea

Tramways were built to join industries to canals. This is the track of the Penydarren tramway; it was along this line that Trevithick demonstrated his steam locomotive.

of anything as vulgar as a canal intruding on their lives was unthinkable. Rennie pacified them by setting part of the canal in short tunnels, decorated by suitably classical motifs, and where it did appear in view it was crossed by ornate iron bridges. But his work is seen at its finest outside the city. Rennie had chosen a line along the north side of the Avon valley, but as the valley closed in, the opposite bank offered a far easier route. An aqueduct was needed to make the crossing and Dundas aqueduct brought an echo of Bath sophistication to the countryside. It is built of Bath stone, a form of limestone that lends itself admirably to decoration. Here it gets the full range of architectural embellishment – pilasters on the piers and a dentilled cornice topped by a balustrade. It is not simply the detailing that makes it so attractive, but it also has that sense of perfect proportion that distinguishes so much Georgian architecture. It is interesting to compare it with his other great aqueduct that carries the Lancaster Canal across the River Lune. This is a far larger structure, carrying the canal 60ft above the river on five semicircular arches. It uses the same classical language as Dundas, but here is treated far more robustly. The stones are no longer smoothed to give an urbane face to the structure, but are left rough hewn and bulky. If Dundas seems perfect as a reminder of the spa city of Bath, then this is no less appropriate of a city linked to the industrial heart of Lancashire. Looking at these two structures, it is not difficult to see how Rennie's later reputation came to depend on important bridges, including London Bridge, which, bizarrely, can now be seen re-erected in Arizona.

One other engineer deserves special mention, though his achievements in extending the canal system were ultimately to lead to its first great challenge. Benjamin Outram was born in Derbyshire in 1764, where his father was a surveyor. At the age of twenty-four he was taken on by Jessop as an assistant on the Cromford Canal. While Jessop struggled with collapsing aqueducts, Outram was put in charge of the 2,966 yard-long Butterley tunnel. It was the longest yet attempted and a somewhat mean affair, with an unusual cross-section, just 9ft wide and only 8ft high, from the bottom of the shallow invert arch to the top of the crown. There was no towpath, and travelling through it must have been like legging down a drain. Nevertheless, the work went well and the only reason it has not survived is due to undermining by a nearby colliery.

There was a fortunate corollary to this venture. During the excavation, Outram found substantial ore deposits, including good quality iron ore. When nearby Butterley Hall was put up for sale the canal company secretary bought the land and leased it to Outram, who set up his own iron works. Later Jessop joined him in the venture, and after Outram's death it became the Butterley ironworks, and it is still at work – and still contributing to the canal world (see p.167).

Outram soon established an interest in making rails for tramways. These were similar to plateways, railways along which the trucks were hauled by horses. He favoured L-shaped rails and actively promoted their use. Their first appearance was on a tramway linking the quarries at Crich to the Cromford Canal. There is a popular story that the name 'tramway' derived from Outram's name, but although it would not be inappropriate it is not true. Just to make the whole thing more confusing, Crich quarry is now home to the National Tramway Museum, which celebrates electric passenger trams, not horse-drawn goods wagons. In the meantime Outram continued to work on a number of important canal schemes, including the Huddersfield Narrow, where Standedge tunnel proved infinitely more difficult than his original Derbyshire enterprise.

He also worked on the Derby Canal, where he needed to build a short aqueduct in a very confined space. He took the obvious decision for someone who was partner in an ironworks to construct it out of cast iron. It was only 44ft long so it could be easily prefabricated and brought to the site. As a result, although it was begun after Telford's Longdon aqueduct, it was the first to be opened by just one month. We know very little about it as it was demolished in 1971 and passed to the care of the local authority who, amazingly, contrived to lose the world's first cast-iron aqueduct.

In 1794 he was appointed chief engineer to the Peak Forest Canal, which was designed mainly to carry stone from extensive quarries in Derbyshire. Outram did not attempt to use cast iron again when it came to building an aqueduct across the River Goyt at Marple, but he did borrow a new technology from another engineer. In the 1760s, William Edwards had built a 140ft-span bridge across the Taff at Pontypridd. An early version had collapsed under the weight of its own masonry. Edwards found an ingenious way of lightening the load, by piercing the spandrils – the stonework between the abutments and the arch. This reduced the load, without weakening the arch itself. Outram used this at Marple. The real importance of this canal, however, can be found in its many tramway connections. The canal itself ended at Buxworth basin, 6 miles short of the main quarries, and was continued by tramway. The basin is now one of the best surviving examples of an Outram canal-tramway interchange to survive. The canal was later to have another interchange at Whaley Bridge, when the Peak Forest was linked to the Cromford Canal by an even longer tramway, the Cromford & High Peak Railway. At Whaley Bridge itself, the canal runs into one end of a substantial wharf building and rails emerge at the far end.

Not everyone was enthusiastic about the new tramways. The Duke of Bridgewater famously remarked that canals would do well enough if they could keep clear of 'those damned tramways'. He saw them as a potential threat and he was quite right to do so, as another Outram scheme would soon demonstrate. Problems were being experienced on the Glamorganshire Canal that served the many iron works around Merthyr Tydfil. There were fifty locks in little over 24 miles and boats were soon queuing to use them. The problem was made far worse by one local ironmaster, Richard Crawshay, who had a controlling interest in the canal. He insisted on his boats having precedence to the understandable annoyance of his rivals, notably Samuel Homfray. The latter decided that he had had enough and commissioned Outram to build a tramway from his works at Penydarren to join the canal at Abercynon, missing out the troublesome locks altogether. There was nothing to distinguish this tramway from many others until 1804, when Homfray invited a Cornish engineer to try an experiment on the line. The engineer was Richard Trevithick and the experiment was to replace the horses with a machine, a steam locomotive. It was not an immediate success – the heavy engine broke the brittle cast-iron rails – but it marked the start of the railway age that was to have a profound effect on canal transport. Outram lived just long enough to see this startling development of his tramways; he died in 1805, shortly after his forty-first birthday. Had he not died so young he might well have made a name for himself in this new world, but instead he is remembered in canal histories but scarcely gets a mention in the story of railways.

In this chapter we have looked at canals mainly in terms of famous engineers who we cheerfully say 'built them'. They planned them, they supervised work on them, but the actual physical work of constructing them was the work of others who never became famous, whose names are rarely recalled and usually not even known.

5

MEN AT WORK

THE WORKFORCE of canal construction was like a pyramid with the chief engineer at the apex. Under him were the resident engineer and the secretary. They in turn would have assistant engineers and clerks to help them. These were all directly employed by the company. Below that came the contractors, a link between the company and the rest of the workforce. Moving lower down the ranks were the skilled workmen – masons, carpenters and the like – and at the bottom of the pile was the very substantial base – the men who did the actual physical work of digging the waterway.

The resident engineer had generally moved on from a career as surveyor or master mason and, with the odd exception of men such as Telford, rarely advanced much further in their profession. Unlike the chief engineers they were tied to the one job, which could last for many years, so they had far fewer options open to them. When Archibald Millar applied for the job on the Lancaster Canal in 1793, he sent in his reference. It described him as, 'A person capable of conducting the business of a Canal through, viz., that he is a good engineer, can carry an accurate level, and has a perfect knowledge of cutting, banking, etc, and also that he is a compleat mason'.

He was appointed and soon discovered that other qualities were also required: almost infinite patience and a willingness to spend a good deal of time travelling up and down the countryside from dawn to dusk. What he would not have expected was that the work would not be finished until 1826.

The largest problem that he faced was derived directly from the way the work was organised. The Lancaster Canal was typical in that the actual physical work was let to contractors, who in turn were liable to parcel it out to sub-contractors. In 1796 contracts on the Lancaster varied from men such as Mr Stevens, who employed 152 men, to Pat O'Neil, whose entire workforce consisted of two other navvies. But by far the biggest and most important contract was let to the firm of Pinkerton and Murray worth £52,000, which is a considerable amount of money considering that the budget for the whole canal was only set at just over £400,000. Big contractors were expected to employ and pay the workforce as well as provide all their own basic equipment. Presumably Pat O'Neil bought his own shovels.

The advantages for the Company were obvious. They did not have the responsibility of dealing personally with hundreds of itinerant workers. The only man who would have liked to see a different system was the resident engineer. His job was to see that all the work was thoroughly carried out and everything done by the book. The contractor's main concern was getting the work done as cheaply as possible, to make sure that there was the maximum gap between the contract price and the costs he had to carry. This conflict of interests was to plague both

Navvies at work, sketched by W.H. Pyne at the end of
the eighteenth century.

Archibald Millar and his colleague, the
company secretary Samuel Gregson.
John Rennie wrote to Gregson in
June 1795, 'I am not surprised at
your being out of patience with
P & M. I think it would be a hard
business for Job himself.' Reading
the letter books and minute books of the Lancaster Canal one can see why.

Pinkerton and Murray both came from Scotland, where having heard of the higher wages
being offered for work in England, they put together gangs of men and negotiated on behalf of
the group. In time, they joined forces and built up sufficient capital to buy equipment and act
as full-time employers, not just spokesmen for a gang. They were not noted for being scrupu-
lously honest. Pinkerton put in an early appearance in the canal records when he tried to bribe
an engineer into giving him a contract and was sent packing. Between them they managed
to develop a growing reputation for incompetence, matched by an eagerness to go to law if
anyone tried to make them pay the consequences. The amazing thing is that they managed to
get as much work as they did. The story of events on the Lancaster was all too typical. The first
source of trouble was that they also had contracts on the Leeds & Liverpool, and lacked the
resources to do both jobs thoroughly. Millar frequently found himself ordering his assistants to
step in to oversee work that should have been looked after by the contractors. When he did
so Pinkerton and Murray promptly complained that he was usurping their authority. Millar
was having none of that, 'You very well know your attention has been very little employed in
respect to the Masonry, and your only Agent (Mr Tate) if he knew anything about the business
had little time to attend to it, there were none but Workmen to direct, and our Agents were in
fact obliged to Act as Superintendents for you.'

The contractors were always eager to hurry the work along. They received staged payments
as they moved from one section to another, so there was always a temptation to start in on a
new section before completing the last. It was a temptation they seldom resisted. Endless letters were sent cataloguing the shortcomings – spreading men too thinly on the ground, not
putting up fences to protect the workings, not providing the materials needed on site and so
on and so on. In Gregson's words, as a result of their negligence 'the whole Country is laid
open to damage'. One of the worst offences appeared in June 1795. The Act of Parliament quite
specifically laid down what water could be used for supplying the canal and what could not.
This was an important point for landowners, big and small, along the line of the route – and it
was especially important to the company when the landowner was a man of importance in the
community. Millar discovered to his horror that the contractors were about to divert a brook
contrary to the Act. You can almost hear him spluttering with rage in his letter to Murray,
'For good Gods sake do not be so very crazy as to turn the said brook without having full
liberty from Lord A. Hamilton.' By September of that year the company had had enough. They
cancelled the contracts with Pinkerton and Murray, called in impartial arbitrators to value the
work they had already finished and paid them off. It did not quite end there. Pinkerton tried
to state his case at the Annual General Meeting and Millar had the great satisfaction of recording the outcome, 'The whole of the Proprietors of Lancaster knew Mr P was making false
Statements from their knowledge of the work round Lancaster. No one seconded him … this I
hope will be our closing scene with these worthy Contractors.'

Men at work on a lock; scenes such as this must have been common while the canals were being built.

It was indeed, to Millar and Gregson's great relief. There was one glowing testimonial to Pinkerton, which puts things in a very different light. 'He was too moderate and too gentle to secure his own interest against the falsehood and calumny, with which, little, mean, and envious individuals assailed his character.' It is rather less convincing when you discover that he wrote the eulogy himself. Between them the pair left a trail of botched work up and down the country, which at times was on the brink of being fraudulent. The Birmingham & Fazeley Canal had scarcely been open more than a few months before lock walls began to crumble. It was only then that it was discovered that behind the outer surface of sound brickwork lay an interior that was a jumble of old, used bricks, some whole, some actually broken. It is as well for the canal system as a whole that the majority of contractors were honest and conscientious.

The conflict of interests was never quite resolved, even if it seldom reached such extremes. One way to exact the most from the contractors was to offer bonuses. Jonathan Woodhouse working on Standedge tunnel for the Huddersfield Canal Company was offered a £400 bonus for every month saved, but had to pay a £200 penalty for each month lost. It was an inducement to get things moving, but not necessarily a recipe for getting careful workmanship. In the end the responsibility for assuring quality lay with the resident engineer and his staff.

A good resident engineer was essential to any enterprise and chief engineers tended, not surprisingly, to continue to employ a first class man whenever they found one. Telford's first choice was his old friend Matthew Davidson, a friendship that survived the fact that Telford was devoted to his native land while Davidson loathed it. He 'declared that he would not accept of a seat in Heaven if there was a Scotchman attached to it'. He was an idiosyncratic character, whose great love was books and he was popularly known as the Walking Library. But he was undeniably good at his job. Telford brought him down to work on the Ellesmere and then employed him again on the Caledonian, where he continued to work well despite his constant grumblings about the people, the weather and all things Scottish. He did get some relief when Telford asked him to oversee his greatest project, the construction of the suspension bridge over the Menai Strait. It was men such as Davidson and Millar who kept the works running smoothly, in the often long gaps between the visits from the chief engineer.

The assistant engineers were generally young men at the start of their careers and hoping to gain enough experience to climb the engineering ladder. They gained the experience, but had to work hard for it, and not always in the most comfortable circumstances. Exley was one of Millar's assistants – not near enough to the top of the hierarchy to merit a Christian name in the official records. He was put in charge of overseeing the work on the Lune aqueduct and had to live on site. A sort of oversized hut was built, in which he had a room of his own, but he had to share the kitchen with the man running the steam engine. Unfortunately the engine man, Mr Richards, proved a far from sober companion – 'has kept the can to his head for 3 or 4 days continually'. Millar, rather reluctantly it seems, allowed Exley to have a floor in his room, a luxury not considered necessary for the kitchen, but noted that he had designed everything on 'the most frugal plan'. A good assistant could earn as much as £250 a year, but the prize was in front of his eyes – Rennie got £600 a year for the Lancaster alone, and was only required to attend for five months out of the twelve. It was a great incentive, and must have made sharing a hovel with a drunken engine man just about bearable.

The administration was the responsibility of the secretary. Money was a recurring problem. Investors who bought canal shares in the hopes of making fat profits tended to overlook the responsibilities that they took on at the same time. The Company could call for further funds when needed and they had the legal right to do so, up to a specified limit. This often came as an unwelcome shock to the investors, and one of the secretary's recurring jobs was to wheedle, implore or threaten until they paid up. It was seldom easy. Another major responsibility was agreeing contracts for the land that had to be purchased along the route. Just as investors were unaware of their legal obligations, so too landowners were often astonished to find that once the Act had been passed they had to sell land, whether they wanted to or not. The request by the Oxford Canal Committee for a settlement on land purchase from Lord Spencer brought an agonised reply from his land agent:

> It would seem by what you say that so very considerable a quantity of land as 16 or 17 acres is the object of your attention and so far as I can judge from the complexion of your letter it is not in Lord Spencer's power to refuse the Accommodation however inconvenient or disagreeable it may be to his Lordship to furnish it. I shall be obliged to you to inform me, as I really do not know, if this is the case, and if Lord Spencer has no other privileges or Command over his own property than that of treating for the Price of any such parts of it, much or little, as the Company may happen to discover are convenient for their use.

Lord Spencer was soon to discover that that was indeed the case. His Lordship did, however, have one weapon he could use in retaliation. The Act did not set any prices on the land, only set out that if agreement could not be reached, then the matter had to go to arbitration. Until they had the land, then no work could go forward, so the Company always wanted to settle negotiations as quickly as possible. In general, they were prepared to pay a good price rather than see progress on the canal halted, while lawyers argued over a few extra pennies per acre. For the secretary it was a difficult balancing act, which could go either way. Speed was essential. If the Company could start negotiations well in advance of when the land was needed, then they had the edge, and could turn down some of the ludicrously extravagant claims made by some landowners. There were times when the battle was lost altogether.

Lord Anson had a large estate close to the line of the Birmingham & Liverpool Canal. Telford was attempting to keep as straight a line as possible, and the obvious, and easiest, line lay right through the centre of Shelmore Wood, which just happened to be where his Lordship bred his pheasants. The opposition to the direct line was so powerful – and the likely costs of settlement

Pile driving to provide bank protection.

so high – that the Company reluctantly had to bypass the woodland. This involved rather more than a simple detour: the canal had to be carried on a mile-long, high embankment. Work on stabilising the bank proved so difficult that it was not even completed in Telford's lifetime.

The secretary shared with the engineer the task of dealing with the contractors, for example ,measuring the work and assessing its value. This involved regular visits to sites up and down the length of the canal; his was very far from being just a desk-bound occupation. He was also the man who received the complaints of every aggrieved citizen along the line of the canal – workmen had left gates open and allowed cattle to get out, carts had created ruts over their fields and so on. He was even required to deal out justice. The secretary on the Coventry Canal was ordered to prosecute a workman for the heinous crime of stealing 'part of a plank'. Sadly, there is no record of which particular bit of the item he pinched. It must have been incredibly aggravating to have to spend time on such trivial matters. As work on the canal neared completion, the demands on the engineering staff diminished. The work of the secretary, by contrast, increased. He now had the extra work of dealing with the traffic that was starting to use the waterway. Such a responsible position was not particularly well paid. Gregson received a modest £250 a year from the Lancaster Company, but he had something the engineers lacked. He had security. His job did not come to an end once the canal was open. He could look forward to a future of profitable trade and a calmer life, untroubled by contractors.

There was never a shortage of jobs for skilled workmen in the canals. There were the obvious trades – bricklayers, masons and carpenters. The canal world was largely self-sufficient. The waterways were built precisely because of the wretched nature of transport in the country, so during the period of construction every effort was made to use local materials. When stone was needed, quarries were opened near the site and simple temporary tramways were often laid down to move the material down to the workings. In regions which lacked suitable building

Installing new lock gates using sheer legs.

stone, brick had to be used as a replacement. Clay pits were dug and the bricks were made in situ. To turn a lump of porous clay into a waterproof brick it has to be heated for at least eight to twelve hours at temperatures in the region of 1,000°C and then slowly cooled. In brick-works this would be done in permanent kilns, but at a canalside site it was more likely to be carried out in clamps. These are carefully built up structures of unfired bricks, with a coal fire at the heart. This is cheap to do, but highly inefficient. There is a tendency for the bricks at the centre to get burned and those on the outside never to get hot enough to vitrify correctly. It was, in short, very hit and miss but cheap. One result of this method was that bricks from any one firing came out in a rich variety of colours, from pallid pink to the deepest red. You can see the effects on canal bridges up and down the country, where even the most modest structure has a rich variety of hues and textures that you never get with modern machine-made bricks. Making the bricks was every bit as skilled an occupation as laying them.

Self-sufficiency came down to even the smallest, apparently insignificant items. When canal construction began there were few precedents to build on. On the Coventry Canal, the contractor Thomas Sheasky was a man of considerable means – he held two contracts, one for £7,580 and the other for £9,460 – but even he had to have a barrow sent down from Staffordshire to find out what would be the best type of wheelbarrow for canal work. He then put in an order for 100 barrows at 7s each. After that he sent a carpenter to Birmingham to see a lock being constructed, so that he could find out how to make lock gates. Everything had to be learned, but in time, as the canals spread across the country, such work became a matter of common experience. For most of the skilled workers, however, the work was simply the sort of thing they had always been doing. It made no difference to a mason whether he was building a bridge over a running stream or a still canal, and to him an aqueduct was no more than a big bridge that happened to have boats running across the top instead of horses and carts. If this

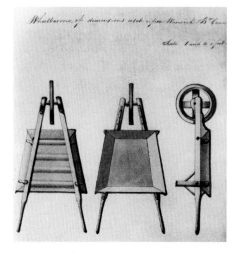

Measured drawing of a wheelbarrow for use in the construction of the Warwick & Birmingham Canal.

was true of the artisans then it should have been equally true of the great numbers of men whose main task was, in effect, to dig a very long, wide, deep ditch. The story is not quite that simple.

When the Duke of Bridgewater undertook the first canal he had no real labour problems; he could call on men from his own estate and miners from his colliery for the bulk of the work. As the canals spread and workforces grew, so the companies, either through their own efforts or through contractors, had to lure manual labourers from other occupations such as farming. The obvious incentive was higher wages, but no one wanted to pay too much. The Committee of the Forth & Clyde Canal were able to report that by 1768, they had over 1,000 men at work and the contractors had managed to keep pay down to tenpence (4p) a day; though they had to pay an extra penny to persuade them to work in winter. The Company took some responsibility for the men and agreed to pay doctors' fees to cover the various injuries they might receive in the work. They assumed the costs would be modest, but when bills came in for the first year totalling £81 they promptly cancelled the arrangement and stopped payment. They decided it was not, after all, anything to do with them – the contractors should pay. This ambivalent attitude towards the workmen was to persist throughout the canal construction period: it seldom operated in favour of the men. One unfortunate side effect for historians is that injuries and deaths were rarely recorded by the company, so we have very little idea how many accidents occurred. One man who did take notice was Thomas Telford, who recorded that only one man died while building Pontcysyllte – and that, he wrote, was his own fault anyway. The fact that he was proud of the safety record suggests it might well have been worse elsewhere. There is nothing in the canal world to match the memorial, in Otley churchyard, to the twenty-three navvies who died building Bramhope tunnel on the Leeds & Thirsk Railway.

Some of those who went to work on the early canals found that life was no harder than it had been down on the farm, but was much better paid. There was an obvious temptation to go and look for another canal to work on, rather than return to the land. They were welcome, for they had become accustomed to the work, and developed strength and skills. The Revd Stebbing Shaw took a trip around Britain and called in to take a look at work on the Greywell tunnel on the Basingstoke Canal. He noted about a hundred men at work near the mouth of the tunnel and asked where they had come from. He discovered that the Company had asked the contractors to recruit locally where possible, but the scheme was not working well. 'Such is the power of use over nature, that while these industrious poor are by all their efforts incapable of earning a sustenance, those who are brought from similar works, cheerfully earn a comfortable support.' A new breed of working men had evolved, and they were given a name. They were navigators, the men who built the navigations, which was a bit of a mouthful, so it was soon simply abbreviated to the familiar navvy.

There is remarkably little information about where the navvies came from. There is a tradition that says that many came from Ireland, which would not be surprising given the widespread poverty in that country at the end of the eighteenth century. This idea is reinforced obliquely by questions that were asked about the workforce on the Caledonian Canal.

One of the main objects of putting money into the venture was to provide work for the local Highlanders, and it was suggested that, in fact, there was a very large Irish contingent at work. Telford produced figures to show that over the period between 1804 and 1812 there was never more than one Irishmen, four Welshmen and one Englishman at work and hundreds of Scots. He did, however, admit that he had brought Highlanders in, who had already worked on canals in England 'to prove useful examples to others who have not been accustomed to that sort of employment'. In general the pay rates were comparatively low on the Caledonian compared to payments in England, averaging 1s 6d (7.5p) per day and when men made their way up north expecting to get their usual rates they were sent packing. Pay could be as high as 2s 6d (12.5p) on English canals and that was their appeal. It sounds precious little to us today, but it has to be put in context. In 1795, Justices of the Peace met at Speenhamland in Berkshire to discuss how much should be paid out of the poor rate to support impoverished farm workers. They settled on a basic living wage for a married man of 6s (30p) a week, with an extra 1s (5p) a week for each additional child under working age. The Speenhamland system soon spread throughout the country, providing unscrupulous farmers with the opportunity to pay less than 6s a week, safe in the knowledge that the parish would make up the difference. There was no relief of any kind available for single men – if they wanted a living wage they either had to find a wife or leave home to look for a living wage somewhere else. It is little wonder that men chose the latter course, with the prospect of earning at least twice as much money as they would have done down on the farm.

It is frustrating to read accounts of construction scenes on the canals and find so little information about the men who built them. An all too typical example is this description from a visitor to the Bridgewater Canal: 'The whole posse appeared, as I conceive did that of the Tyrians, when they wanted houses to put their heads in, and were building Carthage.' Which is little more than a classical scholar saying that there were a lot of men looking really rather busy. Other travellers found equally unhelpful analogies – ants on an anthill being a popular favourite. The official records are little better. Because the men were employed by the contractors, they only get mentioned at all when things were going wrong. They were anonymous men in an anonymous army. What we do know is that it was an exceptional army of remarkably tough and strong men. Most of us have been out digging at some time or other and have probably complained of backache at the end of the day, but it is doubtful if any of us could shift the earth moved by an experienced navvy. They were not merely turning over topsoil, but were digging deep down into the underlying clay and shovelling it up into barrows. It was estimated that an experienced man could shift 12 cubic yards a day. That's a trench 3ft wide, 3ft deep and 36ft long – no mechanical aids, just shovel and pickaxe. That is a rather rough measure, but if it seems unlikely then it can be set against a much more accurate account from the railway age, and the railway navvies in the early days often arrived straight from working on the canals. The famous contractor, Thomas Brassey, provided details of what he expected of his men. His teams filled wagons, which was even harder work than filling a wheelbarrow, as the earth had to be thrown up to a height of 6ft to clear the sides of the truck – and each man was expected to shift 15cu.ft a day. This was just not just like digging a long trench, it was digging a trench and chucking the soil over a head-high garden wall! It makes the canal navvies' effort seem almost puny.

One of the very few accounts of navvies and their work during this period comes from Canada in the 1820s, with the construction of the Rideau Canal in Ontario. The workforce was largely made up of thousands of Irish Immigrants – estimates of numbers vary from 2,000 to 6,000. They must have been not so very different in experience from the Irish who came to work on English canals, and John MacTaggart, the Clerk of Works, described the chaotic scenes at the workings:

Navvies, some obviously very young, photographed during construction of the Manchester Ship Canal.

> Even in their spade and pickaxe business, the men receive dreadful accidents; as excavating in a wilderness is quite a different thing from doing that kind of labour in a cleared country. Thus they have to pool in, as the tactics of the art go – that is, dig beneath the roots of trees, which not infrequently fall down and smother them.

Others were tempted to take work in the quarries, set up along the line:

> They thought there were good wages for this work, never thinking that they did not understand the business. Of course, many of them were blasted to pieces by their own shots, others killed by the stones falling on them. I have seen heads, arms, and legs, blown in all directions; and it is vain for overseers to warn them of their danger, for they will pay no attention.

It is unlikely that anything quite that dramatic happened on British canals, but it does demonstrate the dangers that would have faced the inexperienced navvies, when they first started work on the canals.

However rough and ready the statistics might be, they do show one thing – the navvies earned their money. Not that their employers always thought so. Requests for higher rates were not encouraged and they represented some of the few occasions when the company officials and contractors acted in complete unanimity. When workers for one contractor put in for extra pay, the word was sent out to other contractors down the line not to take on any who tried to find better rates with them. When news did get out that one contractor was paying slightly more, there was an understandable demand from other navvies for the same rate or at least something near it. When men on the Lancaster heard that others were getting *2s 6d* on

The workforce posing during the repairs to Blisworth tunnel in 1910.

another part of the line for doing the same work, they put in for a weekly wage of 14*s* based on a ten-hour day, with overtime at 3*d* an hour. The canal committee were horrified at such a suggestion, promptly sacked the lot and brought in fresh, cheaper navvies to take their place.

The navvies could get a reasonable wage – reasonable that is in terms of other manual workers. But it was not quite as good as it seemed. They arrived in an area as strangers and were seen as ripe for exploitation. In some cases, as on the Leeds & Liverpool in the 1770s, the situation got so bad that the Company was almost prepared to step in. A proposal was put that they should 'consider if any scheme can be come into for the convenience of such Labourers by erecting Tents, Booths, &c &c providing them with meat and drink at a more easy expence.' There is no record of anything actually being done. In the remote Highlands, where the Caledonian was being built, the Company actually set up a small brewery to provide the men with beer and kept cows for their milk. It was not entirely a charitable scheme: the main object was to persuade the men 'to relinquish the pernicious habit of drinking Whiskey'. Somehow the idea of 1,000 hard-working Scots turning down the dram for half a pint of bitter or a glass of milk seems a touch unrealistic. But beer was, in fact, accepted as an essential for the working man. During heavy work, such as puddling, the contractor would be expected to provide a free allowance. It was not entirely altruistic: beer might make you drunk, but the local water supply could kill you.

The greatest problem many of the men faced came about partly because of a shortage of coin of the realm. This was a time when banks were few and getting enough money together for paydays was often near impossible. On some canals contractors issued the navvies with tokens instead of coins, which could either be used to purchase goods directly from the contractors, or, in theory at least, could be used at the local inns and stores. The contractor promised to change the tokens for their face value in coins after they had been spent. The scheme had quite a lot of benefits. The contractors could guarantee that there would always be sufficient tokens available to meet the demand, so the men always got paid. In practice, however, the men often had trouble using the tokens for anything much. They could only go to establishments that had made arrangements with the contractors and had no choice other than to pay the prices asked. Where the contractors provided the goods themselves, things could be even worse. A man got his pay and, in effect, handed it back again. The contractor got the man's labour and then got a second round of profit by selling him the necessities of life.

Accommodation was also a problem. Men got lodgings where they could, but it was not always easy to find. In remoter areas, the Company put up huts and sheds, and where they weren't provided the navvies erected their own and shanty towns grew up along the line. Some accommodation, however, was a good deal better. The Company required quite a number of buildings, so it made sense to let them out until they were needed. These could be quite grand, particularly where they served major projects and a workforce would be on site for months or even years. One of these was Sapperton tunnel on the Thames & Severn Canal. The Company built barracks at both ends of the tunnel, Daneway and Sapperton, the larger Tunnel House at the former being an imposing three-storey building. Once the canal was opened they both became inns, and so they are still, though Tunnel House lost one storey in a fire. No doubt in later years they did a roaring trade with boatmen recovering from the arduous task of legging through the long tunnel.

Given the treatment they could expect to receive from their employers, the navvies felt they owed little in the way of loyalty to anyone. At harvest time, farmers could seldom get all the help they needed and were prepared to pay good wages to hard workers. Navvies left the diggings, made what cash they could, and then reported back for work several weeks later. The employers moaned and complained, but there was nothing they could do about it. There was also a sort of bush telegraph operating among the men, so that they kept each other informed on the going rates around the country. This was invaluable for men who were 'on the tramp' moving around the country from one set of diggings to another. But no matter where they went they were still part of the same navvy community, where the men helped each other out as and when they could.

If one relied entirely on official documents and newspaper accounts to get a picture of navvy life, what you would end up with would be a horror story of depravity and violence. There is no doubt that these men, working hard, living rough and generally treated as outcasts, were never likely to stick to the rules of polite society. Peter Lecount, writing in 1839, some years after canal construction had mainly ended, summed up the general attitude towards these men:

> These banditti, known in some parts of England by the name of 'Navies' or 'Navigators', and in others by that of 'Bankers', are generally the terror of the surrounding country, they are as completely a class by themselves as the Gipsies. Possessed of all the daring recklessness of the Smuggler, without any of his redeeming qualities, their ferocious behaviour can only be equalled by the brutality of their language. It may be truly said, their hand is against every man, and before they have been long located, every man's hand is against them: and woe betide any woman, with the slightest share of modesty, whose ears they can assail.

It is a terrifying indictment of thousands of men, and one is left wondering just what hideous crimes they committed to justify such words. But, reading the whole piece through, there is only one actual offence listed – using bad and probably suggestive language in front of ladies. Which would put them on a par with a lot of men working on building sites in modern

Britain. Lecount was, in fact, arguing a case for using a different system of employment on the railways and had good reason to turn the navvy into the bogeyman. Nevertheless, there are too many cases of navvy riots for anyone to pretend they were no more than hard-working men, who spent their spare time reading and sipping tea. Often the problem can be traced back to the payment system of using tokens. In theory they could be used as hard cash, but in practice local innkeepers and stores were disinclined to accept the tokens or change them. Resentment would build up, fuelled by drink, and the results could be disastrous. This is just what happened at Sampford Peverell on the Grand Western Canal in Somerset in 1811. The navvies had tried to cash their tokens, but had to settle for taking payment in ale. Resentful and angry, the drinking did nothing to improve their tempers. A group left the local inn and promptly bumped into a man, Mr Chave, who had been involved in a notorious fraud. That was sufficient for the navvies to follow him, taunting him and generally appearing threatening. Chave scurried home and tried to scare off the mob that gathered outside by firing a pistol in their general direction. It hit one of the men who died on the spot. That triggered a full-scale riot, during which one of Chave's employees was savagely beaten. The local press was indignant and demanded that the rioters should be brought to justice, though they chose not to mention that it was Chave who had used a firearm and the only man to die was a navvy. The story illustrates, however, two things: that the navvies certainly could be easily turned into a mob, especially after heavy drinking, and that whatever the actual circumstances, they could expect to take all the blame.

There are quite enough stories about riots to prove that the navvies were not blameless. It is often difficult, however, from the published reports to get a clear picture of what actually happened. Here is a typical newspaper report, describing events in Nottingham in July 1793:

> A scandalous riot took place in this town on Wednesday evening, and continued till midnight … A number of navigators, disorderly boys and others, were drawn together by the music of some recruiting parties, who were parading as usual in the evening. After various proceedings, which manifested a mischievous disposition… They attacked the Plough Ale-house, the Sun, the houses of Mr. Haywood, a plumber and glazier, of Mr. Keyworth, a respectable maltster, in the Long-row, and of Mr. Homer, ironmonger … At length these daring miscreants proceeded to the house of Mr. Goldknow, our much respected Mayor, and the attack becoming very violent, his family were under the unpleasant necessity of defending the house with fire arms. They fired upon the rioters and several were dangerously wounded, one or two of whom, we hear, is since dead. The mob, upon this spirited proceeding, dispersed.

What is notably lacking from this report is even a hint as to how the riot started in the first place. Once again, the only fatalities were among the navvies themselves. The navvy riots have to be put in the context of an often violent age. The Church and King riots of 1791, when supporters of the French Revolution, real or alleged, were attacked, caused far more damage than any navvy riot. Among those who suffered was the famous chemist Joseph Priestley, whose house was burned with all his scientific papers and he was forced to leave the country. Elections were also notoriously violent times, and one election at least enabled the navvies to cash in and profit from their reputation for violence. Men left the workings on the Lancaster Canal, lured away by lucrative offers from the supporters of Lord Stanley, which as the Canal Company minutes note 'can be for no other purpose than to riot and do mischief.' Virtuous citizens, it seems, were not above treating navvies as a convenient rent-a-mob.

No one can pretend that the navvies were blameless citizens, but one has to remember that no newspapers ever ran stories reporting: 'Lots of men at work, nothing special happened.' Yet that was surely the reality. Most of the time these men stayed at work, performing tasks that would be far beyond the physical capabilities of most citizens. When they took time off, it is very likely that many of them drank too much and got rowdy and that sometimes this spilled over into more serious rioting. But the sparsity of reports suggests that this was exceptional not the commonplace that some of their contemporaries suggested. Perhaps the most telling comment comes in Lecount's comparison of navvies and gypsies. These were communities living outside the settled world and therefore not to be trusted. They were people to be kept at arms' length and even if they weren't actually doing anything wrong it was as well to be aware that they probably would be doing so quite soon. It was an attitude that was later to be applied to the boat families, who would take over the role of potential villains once trade started to flow along the new waterways. But before that happened, there was to be the one occasion when all involved in the construction could get together to enjoy themselves: the official opening.

The opening was always a great occasion. There would be a procession of boats, usually with the proprietors at the front, bands would play and cannon would fire a salute. Usually all went well, though at the opening of the Gloucester & Berkeley Canal three local enthusiasts tried to join in the fun by hauling out an ancient cannon and lighting it – it blew up. There would inevitably be speeches and a local amateur poet could usually be relied upon to provide the words for a celebratory song of usually dubious quality. One local wrote nineteen excruciating verses in praise of George Tennant, chief promoter of the Neath & Swansea Junction Canal, which ended with these lines:

> I hope when he's dead and laid in his grave,
> His soul will in heaven be eternally saved;
> It will then be recorded for ages to tell,
> Who was the great founder of Neath Junction Canal.

William McGonagall himself could not have done better.

The canal was, at least, in its day, a highly successful enterprise. The same cannot be said for the next canal celebrated in a song, suggesting it was one of the most magnificent works in the kingdom:

> All hail this grand day when with gay colours flying,
> The barges are seen on the current to glide
> When with fond emulation all parties are vying
> To make our Canal of Old England the pride.

The sentiments were very suitable for some grand undertaking, but this was the modest Croydon Canal, opened in 1809. It was never very prosperous and by 1836 it had been sold off to a railway company. Still, everyone had a good day out. Even the navvies got to share in the celebrations on most canals. On the opening of the Pontcysyllte aqueduct, the men had a roast ox and plenty of ale, while the rest enjoyed 'a sumptuous feast'. At the end, the proprietors could sit back and wait for trade to start and for their investments to start earning some money. The navvies went their separate ways, looking for work in another part of the system. And the canal itself was open for business.

6

THE WORKING CANAL

WHEN LOOKING at the construction of canals, it is only natural to concentrate on the engineering problems involved in getting boats from one end of the waterway to the other. But there was also a lot of building work going on of a very conventional nature, buildings that were needed to keep the canal functioning once it was open. The grandest were inevitably the company offices, but many have been neglected, no longer needed when canals came under a central organisation. The Oxford Canal Company offices, for example, were left stranded some distance from the canal, when the basin in the city centre was filled in and turned into a car park. They can now be found down a narrow lane, behind Nuffield College. But hunt them out and you will find a typical neoclassical building with a portico embellished with a coat of arms. It shows Britannia holding a shield and behind her the canal can be seen with the university library, the Radcliffe Camera, in the background. The fact that the canal never went anywhere near that building was not considered important. Even more forlorn are the offices of the Stroudwater Canal, again built in a typically classical style, but now almost forgotten, lurking behind a busy roundabout on the main road out of Stroud. Both buildings, however, send out the same message; this is an important concern, dignified and reliable. Inside were the sort of offices any administration might have, where clerks scratched away at ledgers and accountants tried to balance the books. Because, in this at least, a canal company was no different from any other big organisation, then or now: it was there to make a profit. It was how the profits were made that set them apart. The other structures scattered along the length of the waterway tell the rest of the story: lock cottages, toll houses, maintenance yards, wharfs and warehouses.

The canal companies rarely ran their own boats, though there were some notable exceptions, such as the Grand Junction, though even they eventually ended the practice. Instead companies relied on tolls paid by every boat that used the waterway. The actual charges were laid down in the original Act of Parliament. Some were very simple. The Act for the Trent & Mersey, for example, specified with unnecessary elaboration that they could charge 1¼d (0.5p) per ton per mile for 'coal, stone, timber and all kinds of goods whatever'. The only exceptions were stone for road building and manure, which were toll free, provided the boats 'pass through the locks at such time as the waste water flows over the weir'. By contrast, the far more modest Pocklington Canal had a scale of charges of staggering complexity. The canal was divided up into twenty sections, with a different rate for each section. The different goods were also split up into twenty five categories, one of which reads: 'boxes, cloth, coffee, dying woods, dry goods, fruit in chests or boxes, glass, groceries, hides, hops, paint, parcels, pitch, rice, saltpetre, spirits, starch, sumach, tar, tea, tin, tobacco, turpentine, wines, welds and yarn.' Pity the poor

The Worcester & Birmingham Canal Company offices at Kings Norton, photographed in the early twentieth century.

A Birmingham toll ticket from 1790.

clerk whose job it was to sort out the charges on a mixed cargo, each item of which had to be allocated one of the 500 available rates. If, like me when I first read the list, you are wondering what 'sumach' might be, it is a plant used in tanning.

The list of tolls tells quite a lot about the canals. The proprietors of the Trent & Mersey knew that their canal was specifically designed for carrying bulk goods such as coal and they weren't too bothered about the details. The Pocklington served rural Yorkshire and had to squeeze out every penny it could from any goods they could think of. The Duke of Bridgewater always maintained that the good canal had coal at the heel of it and the Trent & Mersey matched his criterion and thrived. Not so the Pocklington. It staggered on into the twentieth century and would have closed earlier but for an aggravating clause in the original Act. The canal was used by simple sailing barges, called keels, and as long as any keel was still trading the company was required to keep the canal open. Trade dwindled but refused to vanish altogether, as one stubborn keel owner continued to appear with his vessel. The company finally found a solution in 1932 – they bought him a lorry.

The majority of canals had a sliding scale of tolls, though none were anything like as complex as the Pocklington table of charges. To work out how much a carrier had to pay, the company had to know three things: what the cargo was, how much it weighed and how far along the canal the boat was travelling. The carrier provided all this information in a written note. The question the canal company had to answer was: is the information correct? Is the cargo what it claims to be, is that the correct weight and how far is the boat really going? The first should have been the easiest to answer, but the carriers were not above subterfuge. For example, on the Grand Junction the charge for carrying coal was three times as high as the charge for limestone. Anyone carrying both would be tempted to put the coal at the bottom and cover it with stone, hoping the toll collector would not notice that the boat was unusually high in the water for one filled to the gunwales with heavy stone.

To find the weight of the cargo, each boat using the canal had to be gauged. A new boat was taken to a dock, where the waterline was marked on the hull with the boat empty, then weights were gradually added and the appropriate marks made. Sometimes these were cut into the hull, or they could be engraved on an attached metal plate. More rarely, canal companies had weighing machines, basically no more than overgrown scales. The boat was floated on and the water emptied. Such a machine, which was originally used in Cardiff, can now be seen beside the locks at Stoke Bruerne on the Grand Union Canal. A rather more sophisticated system was later introduced in which the information on depths was recorded for each boat and copies kept in the company's toll offices; this prevented unscrupulous owners tampering with the original marks. Such careful measures did not prevent the boatmen from trying a few tricks. A man might suddenly spot an old friend on the opposite bank and step to the side of the boat for a chat – that side dipped down and the side being measured rose up, suggesting a lighter load. Keeping records of where the boat went was very much simpler, but it all had to be checked and someone had to do the checking.

Most companies had toll offices, which like the toll houses on turnpike roads of the time, had to provide a view up and down the canal. Some, such as those on the Staffs & Worcester, were quite elaborate affairs. The one at Stourport looks very much like the gatehouse to some grand estate, while further up the canal at The Bratch, the toll house is more like a gazebo in the parkland, a little octagonal building with a conical roof. Most, however, were strictly practical, and none more so than on the improved main line of the Birmingham Canal, where they sit right in the middle of the waterway on little islands. On some canals, however, the job of collecting and regulating tolls fell to the lock keeper. On the Oxford Canal in the middle of the nineteenth century, the Banbury lock keeper had the responsibility of gauging every boat

A typical island tollhouse on the Birmingham Canal with not much traffic in sight.

Tolls were calculated on the basis of the goods carried and their weight. Each boat was graduated to show how deep it was in the water for a particular load. Here the official is gauging the boat, measuring the depth.

passing the lock. This added responsibility earned him a weekly wage of 17s (85p) while on quiet sections of the canal the wage could drop as low as 8s (40p). By the end of the nineteenth century, the job of lock keeper and toll clerk was held by Arthur Thornton, and his daughter, Mrs Rose Macdonald, remembered her life there as a child. She described the toll office in the years near the beginning of the First World War:

> It had a lovely cream coloured rose climbing all over. Very cosy inside, a desk right across the window where he perched on a high stool to write in the big ledger. Two armchairs, one was his, and one for the current dog and probably one of us children, or one of his cronies, a cosy coal fire in cold weather … my brightest memories of him are those of him skipping as lively across the boats, with his gauge over his arm, notebook and pencil at the ready.

The family actually lived in nearby Factory Street, that ran down to the canal, and were part of a wider canal community, including the boat family of Mr and Mrs Alfred Hone, who had saved up and bought a little house 'on the land' and the Tooley family whose traditional boatyard had been turning out narrow boats since the canal had been built. Readers of L.T.C. Rolt's classic book *Narrow Boat* will remember that he made his way down Factory Street, which had a shop on the corner advertising 'Tripe, Ox Heels and Neatsfoot Oil for Sale' and at the end of the street the boat *Cressy* was moored. He had bought it as a horse boat and it was about to be converted into a pleasure boat and have an engine installed at Tooley's boatyard. It is always pleasing to find this sense of continuity, but there is no point in looking for the rose-covered office any more – it was demolished by a bomb in the Second World War.

The rates paid to lock keepers seem very low but most had a rent-free house and usually were given a ration of coal to see them through the winter. Cynics have suggested that the coal ration was simply a way of handing out legitimately what would probably have been taken from passing boats anyway. It is rather reminiscent of a later age, when a fireman whose house backed onto the tracks, reckoned he could get a good half dozen shovelfulls into his garden as he steamed past. In general, the canal companies looked after the lock keepers very well and were aware of the problems that came from living in a tied cottage. In the 1850s ,the keeper at Cropredy lock had died, but his widow was kept on and was still officially in charge of the lock when she was seventy-four years old, though in practice her son did most of the work. Other widows were allowed a small pension and given help with accommodation. It may have been a low pay job, but it had its compensations. But what exactly did lock keepers do for their money?

One thing they did not do was open and close the locks, though they were in charge of them. That was down to the boatmen, who were always in a hurry and more concerned with speed than being careful with company property. John Smeaton, advising on staffing for the Forth & Clyde in 1767, wrote that the main task of the lock keeper was 'to be a check upon the bargemen from doing damage to the works, by running against the lock-gates, leaving the cloughs [sluices] running, so as to let off the water, &c.' In fact, by far the most important responsibility of the lock keeper was managing the water levels along the canal. A dried-up pound could bring the whole system to a halt, and during a hot summer water conservation was the overriding concern. Winter brought a different, but no less pressing problem. Lock keepers would often be up before dawn, breaking the ice in a lock and freeing the gates from its steely grip. It was beyond the resources of an individual to keep a long pound ice-free, but he could call on the services of an ice-breaker. This was a heavy boat, its hull reinforced with iron. Along the centre line was a rail that men could cling to. The ice breaker would be towed as fast as possible towards a frozen section, so that it overrode the ice, and men, standing either side of the rail would rock the boat backwards and forwards to smash the ice. Too little rocking and nothing much was achieved; too much and the men on the boat could end up with a very cold early morning bath. But this was important work. If the ice took hold, the whole canal could be brought to a standstill. It was bad enough for the company, who lost revenue. It was even worse for the boatmen. Those employed by carriers were put on half pay. Those who worked for themselves got nothing, and were even reduced to begging in the streets. An early photograph shows men doing just that, holding a model of a narrow boat in a block of ice to remind passers by of their desperate situation.

The lock keeper was also expected to control the movement of boats. On many canals, some boats had precedence over others at the locks. Cargo boats for example were usually required to give way to passenger boats. This made sense, because the passenger boats had a timetable to keep – sense, that is, to the people who ran those boats. It seemed less satisfactory to a boatman who got paid a fixed fee for a specific journey and for whom, literally, time meant money. It can never have been easy keeping the peace and enforcing the rules. Other rules often dictated that boats could not pass locks at night, but again the boatman in a hurry would try and sneak past in the darkness if he thought he could get away with it. A lock keeper on the Leeds & Liverpool told me of his predecessor who had tried to overcome this problem by an ingenious array of bells fastened with strings to the lock gates. The boatmen got round that one. They employed an army of small boys to ring the bells right through the night. After scores of false alarms, the system was abandoned. Arthur Thornton at Banbury had a much easier life; there was a low drawbridge right next to the lock, which was padlocked each night.

It is no wonder that companies looked after their lock keepers well, for ultimately the smooth running of the whole system depended on them. The keeper who took a pride in his locks was, and is, invaluable. Even those of us who only boat for pleasure know the delight in finding paddle gear well maintained, greased and smoothly running – and have experienced the frustration of coping when it is not. The army of lock keepers has all but disappeared, and the cottages themselves are sometimes no more than holiday homes. So it is easy to think of living in one and working on the canal as quite idyllic. But those lock keepers' duties have to be put in the context of the traffic they were controlling, with literally thousands of boats a year passing up and down the canal, in all seasons and all weathers. There was no time off, no release – as long as the boats came the lock keeper had to stay on the job.

Tolls were not the only source of income for the company. Goods often needed to be stored at wharfs and docks, either waiting to be moved on to their final destination or having been brought by road to be loaded onto boats. Charges were made for storage, and these were much more flexible than tolls. The Trent & Mersey Act remains a model of simplicity, simply saying: 'Wharfage rates for goods remaining more than twenty-four hours, as may be agreed.' A wharf could be anything from a length of canal that had stone coping and mooring rings to make loading and unloading simple to a major complex with secure warehouses. These buildings could vary from plain, single-storey structures in country areas to the immense blocks of a major centre such as Gloucester Docks. The more sophisticated warehouses were built with arches at water level, so that boats could be floated right underneath the building to be handled in the dry. There are examples all over the system, though the most impressive of all, the complex at Ellesmere Port, was mostly destroyed in a major fire. Probably the best surviving complex is at the canal basin in Sheffield, now known as Victoria Quay. Quite why the basin should be renamed is a mystery, and why it should be named after a monarch who came to the throne two decades after its construction is even more mysterious.

The most important sites were at interchanges, particularly those between rivers and canals, where cargoes had to be exchanged between barges and narrow boats. One of the earliest

A Thames lock keeper.

was at Stourport, where the trows and barges of the Severn met the narrow boats of the Staffs & Worcester. The large dock itself could be approached from the river either through a wide barge lock or the conventional narrow lock. Here an impressive array of warehouses and offices was built and a hotel was provided for merchants who came to do business, and in time a whole new town developed around the canal complex. Stourport is, in its way, as typical a Georgian town as Bath or Cheltenham.

One measure of the importance of these docks is the amount of money that had to be expended in creating them – and the huge amount of work involved. The Regent's Canal Dock was designed to link the canal with the tidal Thames at Limehouse. This part of the river, below London Bridge was accessible to the largest sea-going cargo vessels afloat, but the builders had their eyes set on a less exotic trade, the busy colliers that came down the coast from Newcastle. The advantages of having ships brought into a closed dock where they could be loaded and unloaded at all states of the tide were immense, so everything had to be on a suitable scale, with a big ship lock and a deep basin. The original estimate was for nearly £30,000, which can be compared with the estimate for the canal itself, which was only £300,000 for a broad water-way, 8½ miles long, with twelve wide locks and two tunnels. The work was prodigious with the contractor estimating that to build the basin and the ship lock would involve a total excavation of 131,465 cubic yards. That is an awful lot of muck to shift. It was well worth the effort; by the middle of the nineteenth century as many as fifty vessels were entering or leaving the ship lock on every tide. As the cargo of even one ship was enough to keep a whole fleet of narrow boats in business, and the rates per ton per day for storing goods were comparable to the rates per mile for carrying them, this was a lot of business for the canal company. The final cost came out at almost £50,000 but it was money well spent.

Whether it was a major dock servicing ships and barges or a quiet rural wharf, there was one problem that always had to be faced – keeping the stored goods secure. A certain amount of minor pilfering might be tolerated, but large-scale theft was a very different matter. It was well known in Wigan that one particular shop was entirely stocked with goods stolen from the nearby Leeds & Liverpool wharf. The Regent's Canal Company, faced with a bill of almost £6,000 to make the dock area secure by enclosing it behind high walls, opted for the cheaper solution of appointing their own police officer. Few sites required that level of supervision, and on most the job of looking after the goods fell to the wharfinger, with or without staff to help him. This was an important post and was generally better paid than lock keeping.

Trying to keep thieving hands off the goods was only a small part of the wharfinger's duty. Like the lock keeper, he needed to get boats moving in an orderly and efficient manner. At a big wharf, he would need to ensure that priorities were maintained and that loading and unloading was carried out efficiently. As always the boatmen wanted to get away from the wharf as soon as possible – it was dead time as far as they were concerned. The wharfinger had to make firm rules about priorities. His word was law, but the one thing he never did was

Bulbourne maintenance yard on the Grand Union.

Making new lock gates at Bulbourne.

A lot of maintenance work was carried out on site; cutting a new lock beam.

try to tell a boatman how to arrange a cargo. No two boats were ever quite the same, and the boatman knew just how to adjust the loads to create exactly the right balance to make the boat run smoothly through the water – 'make it swim' as they said. The wharfingers had another important role along the canal. Wharf managers also decided which boats took which cargoes. Boatmen soon learned to treat the managers with respect. As a manager for the well-known carrying firm of Fellows, Morton & Clayton, describing his job in the 1920s, explained, 'If a man came to my counter for a job and he'd got a cigarette in his ear, he wouldn't get a job. And if I asked him in the office and he didn't take his cap off he didn't get the job'. On the other hand, the managers knew the communities among whom they worked. They would try and juggle schedules so that a family could be at a certain spot on the system on a certain day if there was a big event such as a marriage or a funeral. They were an invaluable part of the communication system for the floating population.

Canals are no different from any other transport system – they need to be looked after and kept in good repair. Anyone who has ever walked beside an abandoned canal can see how quickly deterioration sets in. To keep everything in good order required a considerable workforce. At the heart of the system was the maintenance yard, which could often contain surprisingly ornate buildings, because, like the head office, they were the public face of the canal. Hartshill yard on the Coventry Canal is a good example of how basic shapes are determined by function but can also be put to decorative use. Curves predominate, simply because boats coming into the yard need to turn quite sharply off the main channel, so walls are rounded to minimise damage. A bow hitting the side would simply slide round rather than coming to a bruising, brick-smashing halt. The motif of curves is then taken up in the main building, with round headed windows matched by the blind arches that decorate the walls. The rather ornate clock tower has a dual function. It is obviously valuable for a community where watches were the preserve of the wealthy, but its louvred sides also provide ventilation for the workshop underneath. One of the stranger examples can be found at Ellesmere, which was a busy little spot. Here the canal company had their offices in the formal, classical, bow-fronted Beech House. The yard by contrast has an exposed wooden frame, similar to the style of many 'black and white' houses for which Shropshire and the surrounding area is famous. It is perhaps unfortunate that today it has rather more the air of suburban mock Tudor.

Inside any of these yards there would be a whole range of activities. Carpenters would be at work making anything that was needed in wood, but the main job was making lock gates – a never-ending task as the gates endured a short life and a brutal one. Leaking gates meant lost time in filling a lock and, rather more importantly lost water. Regular replacement was a necessity and an economy. A blacksmith would be responsible for any ironwork that was required, as well as looking after the horses that would be needed to move material around the system. A mason might also be employed, though it was more likely that, as with bricklaying, the work would go to local men, as and when they were needed, rather than employees of the company.

The main work was out on the canal itself. An engineer would normally be employed to be in overall charge and the canal was split into sections, each looked after by a foreman and labourers. It was a scene of constant activity – keeping the towpath in repair, trimming hedges, cutting weeds, repairing fences and most importantly keeping the channel itself in good order. Dredging by hand was tough work. The commonest method was to use a bag and spoon dredger. This usually consisted of a metal ring, 2–3ft in diameter – the spoon – to which a leather bag was attached. This contraption was then fastened to the end of a large pole. A barge would be moored in the middle of the canal and the spoon dropped to the bottom of the canal, where it sank into the mud. It was dragged along the side of the barge, then hauled out and emptied. It was very hard work and usually needed two men, but life could be made easier if

A portable steam engine has been brought in to work a pump to drain Foxton locks for maintenance.

there was a winch on the barge to help with the dragging. A spoon and bag dredger powered by a Boulton & Watt steam engine was first used for dredging Sunderland harbour, but it was a cumbersome device, and was soon abandoned as a design in favour of the ladder dredger.

The 'ladder' is a continuous belt to which metal buckets are attached. In operation, the ladder is lowered to the bed of the waterway, and once it reaches the bottom, the ladder is rotated, the buckets scoop up the mud, and at the top of the climb they are inverted and emptied. William Jessop designed a dredger that was built at the Butterley works and sent up for use during the construction of the Caledonian Canal. Robert Southey was most impressed when he visited the workings at Fort Augustus with Telford:

> The dredging machine was in action, revolving round and round, and bringing up at every turn matter which had never before been brought up to the air and light. Its chimney poured forth volumes of black smoke, which there was no annoyance in beholding, because there was room enough for it in this wide clear atmosphere.

Modern-day greens might be less enthusiastic. Machines like this have continued in use on wide canals right up to modern times. There is a splendid working example at the waterways museum at Gloucester Docks, which was used both here and along the Gloucester & Sharpness Canal. It was built in Holland in 1925 and when it arrived it came complete with two pairs of clogs, one for the engine driver and the other for the fireman, to protect their feet from the hot, oily metal floor: thoughtful people, the Dutch. Although powered machines were available for major waterways such as the ship canals, on most of the system work continued to be done by hand.

Many of the structures built for the running of the canal still remain, but seldom serving the same purpose. They can act as reminders of what the canals were like in their heyday. For over a hundred years the principle means of hauling boats was literally by horse power.

Stables were very important. Some companies provided buildings that would grace any modern riding school. The stables beside the Chester Canal at Bunbury locks are particularly fine, whereas horses on the Staffs & Worcester at Cookley had to make do with a cave carved out of the soft sandstone cliff. One other source of stabling was always possible, as it catered just as well for the human as the equine taste – the canalside inn or pub.

The canalside pub was more than just a place to stop for a drink at the end of the day. As well as providing stabling, a pub would often have a shop attached, selling basic provisions. For the canal population it was the most important social centre they had. Some pubs would have been very basic. A rare survivor from those days was still open until comparatively recently. The Bird in Hand at Kent Green on the Macclesfield looked from the outside just like an ordinary house. When I first went in there I thought I had actually made a mistake. I had walked into a living room and there was an elderly lady standing in her kitchen door. I was assured that this was indeed the pub, so I ordered a pint. She took a jug, disappeared down the cellar and came back with it full of beer. And if you wanted anything different then that was really hard luck – this was one of those rare places, a pub with a license to sell beer but not wines or spirits. The only indication that it was a pub was the dartboard hung over the fireplace, and when you tried to play on a winter evening it was an interesting exercise in aerodynamics as the dart floated upwards on the hot air from the fire. It was all too clear that it was never going to last, and it hasn't, but it was a reminder that once very many ale houses were just like this. .

Many of the old pubs have an interesting history. The Bridge Inn, at Etruria on the Trent & Mersey, was part of a terrace actually built by Josiah Wedgwood to house workers at his pottery next door. In the nineteenth century, the end house was bought by a former boatman, who decided he could make a good living selling beer to his old workmates. He was very successful and eventually bought up the house next door as well and the result was a fine local with wood panelling, engraved glass and ornate ceramic pump handles. Sadly it was demolished to make way for a road widening scheme; even more sadly the scheme was then cancelled, but too late to save the pub. Happily many pubs still survive, generally offering a warm welcome to the holiday boaters who have taken over from their working predecessors. Not always, however; I did visit one pub on the Macclesfield announcing that anyone coming off a boat was relegated to the public bar and must never disturb the sanctity of the lounge bar where respectable people met.

The canal system was built for boats, but it needed a lot of people tied to the land to keep it viable. Smeaton gave his estimate of who was needed to keep the Forth & Clyde Canal up and running. They were: one toll clerk, two toll collectors, one surveyor, two overseers, two masons, two carpenters, sixteen labourers and six lock keepers. It was probably an underestimate. Moving forward half a century, the employment roll for the Regent's Canal dock listed the following: dock master, assistant dock master, police officer and wharfinger, carpenter, foreman, police officer, night watchman, pier head watchman, eight gate labourers and a writing assistant. The presence of so many gate labourers is a reminder that this was still a world where manual labour was the norm. The ship lock allowed vessels of 60ft beam and 20ft draught, so one can easily imagine the size of the gates and the effort required to open and close them by nothing more than muscle power and a simple capstan, not to mention the effort needed to manhandle a ship through. The canal world was not quite so simple as it might seem, needing a lot of people with a great many different skills. In the end, however, they were all there to serve just one purpose – to get the boats through.

7

THE CRAFT OF THE WATERWAYS

B OATS HAVE been plying the rivers of Britain for literally thousands of years. The earliest could have been no more than hollowed out logs, but the remains of vessels constructed out of planks were discovered in the mud of the Humber and have been dated to around about 1500BC. So by the time the Bridgewater Canal was opened, boat builders had centuries of experience to draw upon. During that time, boat builders had adapted their designs to the particular nature of the waterways on which their craft would trade. A Humber keel looks nothing like a Thames barge, a Norfolk wherry has no more than a vague resemblance to a Severn trow. The canal engineers had to make the decision, whether to adapt their canals to take the existing vessels, as Smeaton did on the Forth & Clyde, or build the canal in the way that seemed feasible and economical and let others build boats to fit it, as Brindley did throughout the English Midlands. As boats were being built for use on rivers long before the first canals appeared, it is only reasonable to start with them. And it seems equally reasonable to start by looking at the Thames. It was Britain's most important trading river, only challenged by the Severn, and it was also the most described, both in words and pictures, so that we have a good idea of what the craft looked like from quite an early date.

The name 'Thames barge' conjures up an instant image of a splendid vessel, with its red mainsail extended from the mast by a long pole or 'sprit', hence the name spritsail barge. It will have a small mizzen mast, a top sail above the main and the bigger barges will have a bowsprit and foresails. These majestic craft could and did make journeys outside the reaches of the river, trading all round the south-east coast of England. Barges like these, however, were scarcely to be seen when the canal age began, and the few that had been built were modest in size and considered rather outrageously modern. Most of the craft then using the river had changed very little since the seventeenth century, and we know quite a lot about them.

Paintings of the time, showing places of interest along the Thames, often include barges as part of the busy life of the river. These vessels are little better than oblong boxes, with the bows sloping down to the waterline, in what is known as a 'swim', rather like oversized punts. There is a single mast set well forward, carrying a square sail suspended from a yardarm. This is just about as simple a vessel as you can imagine. It is the sort of thing a small child could make with a matchbox, a match and a little square of paper. In the paintings they always seem to be bowling along quite merrily, but with this simple rig they had to rely on a following wind to make headway, and the task of moving the vessel was doubly difficult going upstream against the current. Often there was no option but to pull the barge along from the towpath. For most

An oil painting of Humber keels at Selby on the River Ouse. Keels traded as far inland as Sheffield.

of the time, the barges literally depended on manpower. John Taylor, known as the waterman poet, wrote a verse description of the Thames in 1632, in which he made a powerful plea for using the river instead of the roads for transport. He pointed out that eight men pulling a barge could do the work of forty carts or 'waines':

> And every waine to draw them horses five,
> And each two men or boyes to guide or drive,
> Charge of an hundred horse and 80 men
> With eight mens labour would be served then,
> Thus men would be employed, and horse preserv'd
> And all the Countrey at cheape rates be serv'd.

The barges he described carried no more than 20 tons, but by the eighteenth century far bigger vessels were in use, and gangs of fifty or so men were needed to haul them, Frank G.G. Carr, author of the standard work on Thames barges, described these men as being 'of the worst possible character, and a terror to the whole neighbourhood'. One suspects that, rather like the navvies, their evil character was greatly exaggerated.

This type of simple, single-masted sailing barge did not survive on the Thames, but a close relation remained in trade in the north-east until the middle of the twentieth century. The Humber keel was a more sophisticated version of the early Thames barges, and could be thought of as a direct link to medieval ships. Instead of the square ends, the bow and stern are rounded, so that the overall shape of the hull is more date box than matchbox. The tall mast carries a topsail above the square mainsail and manoeuvrability is greatly improved by the use of lee boards. This is a vessel equally at home in the broad waters of the Humber estuary as in the narrow confines of a canal. Seagoing vessels are prevented from slipping sideways and stabilised by means of the keel running along the bottom of the hull. But deep keels are impractical for vessels using the shallow waters of a canal, so the heavy lee board is lowered over the leeside into the water to do the same job. The keels can sail surprisingly close to the wind, by pulling the sail so that it lies at an acute angle to the centre line – sailing vessels do not need to have the wind directly behind them to make progress.

One of the last of the keels, *Comrade*, was owned by Fred Schofield. His first trip on a keel was with his father in 1906 – he was three weeks old at the time. By the 1970s, the trade for sailing barges was virtually over, but *Comrade* was spared the fate of other keels, which were either scrapped or converted to houseboats. Fred agreed to sell her to the Humber Keel and Sloop Preservation Society and stayed on as sailing master. I was lucky enough to sail with him on a few occasions and in his skilful hands the vessel moved as elegantly as any yacht. It was a privilege that cannot be repeated. A keel skipper, in the working days, would be out in all weathers and all conditions, with just himself and a mate to handle the vessel, and he gained the sort of knowledge that only comes with that sort of experience. The Society was incredibly lucky to have Fred to pass on his skills to a new generation so that *Comrade* sails on, and hopefully will do so for many years to come.

Comrade is what is known as a Sheffield boat, which means that she could trade over rivers and canals as far inland as the heart of Sheffield. The limiting factor as always was the size of the smallest lock on the system, which in the case of the Sheffield & South Yorkshire system meant vessels up to 61ft 6in long, 15ft 6in beam, 6ft draught and 10ft headroom. Vessels this size could load up to 100 tons. The interesting statistic is headroom. One look at *Comrade* and it is immediately clear that the mast is well over 10ft high. It needed to be tall to catch the breeze blowing above the tops of canalside trees. But it also had to fit under bridges, which means that it had to be lowered when necessary. Now the very simple rig makes sense; if your mast is going up and down you don't want to deal with a complex of sails and lines each time you move it. The keel may look like an ungainly vessel, but she is beautifully adapted to the job she had to do. This is true of all the working barges; they may follow a basic pattern, but that pattern is dictated by necessity.

Comrade has a sister ship, the Humber sloop *Amy Howson*. The basic hull is the same, but she has a fore and aft rig; in other words, instead of having a sail hung from a yardarm at right angles

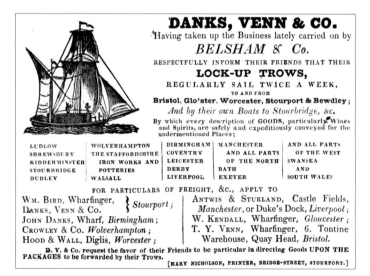

A Severn trow featured in a trade advertisement.

to the centre line of the hull from bow to stern, the mainsail is suspended from a gaff aligned down the centre line, and the whole of the mainsail is always to the rear of the mast. The rig is completed with the addition of a foresail, in front of the mast. Sloops are said to be easier to handle on the tideway and on the lower reaches of the rivers. They regularly traded on the Market Weighton canal, so they needed to be slightly narrower in the beam than the Sheffield keels and of shallower draught. Sloop or keel there were limits to the use of sail power – no one can make progress against a head wind, so both had to be towed when necessary. This was the work of the horse marines. Fred Schofield described these men as a tough breed, who would get up at five in the morning, driving their horses all day and covering as much as 20 miles with a loaded keel.

The square-rigged keel is an exception among the sailing barges that travelled the inland waterways. The majority were variations on the sloop theme of fore and aft rig. The Severn trow began as a square-rigged vessel, not unlike the keel, but in later years was developed with ever more complex rigging. The biggest trows were one or two masted, carrying a topsail above the mainsail and two or more foresails. They were distinguished by the D-shaped transom stern. In later years they regularly traded down the Bristol Channel to South Wales, which in a vessel with an open hold must have been a more than interesting experience in rough weather. The last preserved example, the single-masted *Spry* now has a permanent home at the Ironbridge Gorge Museum. I find it sad to see this handsome vessel removed from her natural element: she was given a brief outing under sail after restoration and was a glorious sight.

Looking across the country to East Anglia one finds a very different version of the sailing barge, adapted for use on both the rivers and the lakes that make up the area of The Broads, the Norfolk wherry. This is another gaff-rigged vessel, with the single mast set very close to the bows, and the large sail has a high peak to make the most of the available wind. It is a very easily managed craft, quite capable of tacking even down the comparatively narrow river reaches, though it really comes into its own on the wide waters of the Broads themselves. It is no wonder that the basic design was adapted to create the wherry yacht. Like most of the sailing barges, there is a small, cosy cabin in the stern for the two-man crew.

Albion has been preserved just as she was in her working days, though now the hatches cover sleeping accommodation instead of cargo. Again I have been able to spend many days on board, and one comes to appreciate how well adapted she is to the region, with its very variable winds. Even the lightest breeze is enough to fill the large sail and as the wind freshens, the sail can be easily adjusted. Only the top is fastened to the gaff, so the bottom can be reefed, rolled up and tied with tapes using – what else – reef knots. But even *Albion* cannot sail in a flat calm, and requires a different motive force. The Broads do not have towpaths, so she has to be moved by quanting. The two crewmen have a quant each – a long

A Norfolk wherry, the trading barge of East Anglia, enjoying excellent sailing conditions.

pole like a punt pole – and they spear it down to the bottom of the river, then push on the quant while walking down the side decking on each side of the cargo hatches. There is one other interesting feature of the wherry. The mast has to be lowered for bridges, but the effort goes into pulling it down, not heaving it up. It has a heavy counterweight at the foot, below decks and once released it rises smoothly back into position.

Once you start looking at these traditional barges you begin to realise what a complex subject it is and just how many variations are possible. It is easy to see how the canal boats of the broad canals are no more than the old river barges, stripped of their sails and rigging and adapted for towing by horses. Many of the original barges were developed to coincide with the improvement in river navigations in the eighteenth century, in the years leading up to the canal age. The Mersey flats, for example, first appeared in the 1730s after the Mersey and the Irwell Navigation had been completed, linking Liverpool and Manchester. As these were the vessels that were to use the Bridgewater Canal once it reached Runcorn, they have a special place in the canal story. Sadly the last of the sailing flats was broken up many years ago, but one feature at least seems to have been favoured, when Brindley decided to halve their width and abandon the sail to create the narrow boat. The long curved tiller of the flats was adapted for the new canal boats and remains in use on horse-drawn boats right up to the present day.

Given the rich variety of the sailing barges, the narrow boat might seem comparatively simple, possibly slightly boring, but the story of the narrow boat is far from dull. For most of us these are the quintessential canal boats, so it is worth looking at them in some detail. The earliest narrow boats were constructed entirely of wood. The construction was very basic. At either end of the keel were two stout wooden posts, set at an angle, the stem and the stern post. The bottom was flat, the bow and stern pointed. From the high stem post, protected by a fender, the sides curved outwards to a point about 7ft from the bows, and this forward area was decked over to provide a storage space. The sides then remained parallel until they curved in again to the stern. At the front of the hold was a triangular board, rising above the fore deck, known as the cratch. At the stern end was the back cabin. The space in between was covered by the top planks, running from the cabin end to the top of the cratch. The end nearer the cratch was covered by tarpaulin and, if a cargo had to be kept dry, the whole of the hold could be covered by sheets fastened over the top plank and tied to the gunwales. At the stern, the cabin was generally closed off by means of hinged doors and a slide over the roof. The steerer could stand in this convenient space between the back of the cabin and the stern post and in winter could get a little welcome warmth, at least on the lower part of the anatomy. The tiller itself slotted into the stern post and could be reversed. It was curved down in use, but could be pointed to the sky to make for easier access to the cabin when the boat was moored. The heavy wooden rudder was protected by fenders. The other essential was the mast for the tow rope.

At the start of the canal age, it was rare for families to live on the boats, and many had only the most rudimentary cabin or none at all. Boats such as these continued in use on the Birmingham canal system, where journeys tended to be short. These were day boats, known as joeys, and they often dispensed with cabins altogether. They were not unique. On the canals of South Wales, linking the collieries and ironworks of the valleys to the Bristol Channel ports. distances were also short. The Glamorganshire Canal, from Merthyr Tydfil to Cardiff, for example, was just over 24 miles long, which might have been possible in a day's travel if there hadn't been fifty locks to contend with. The canal did not connect to any larger system, so having reached one end the only thing a boat could do was go back the way it had come. Many of the boats were double ended, so there was no need to find space to turn round – just move the rudder from one end to the other. They very rarely had any cabins. There were some interesting variations on the theme of very basic boats.

A Northwich class narrow boat being launched.

The narrow boat *Unostentatious* receiving its traditional painted decoration.

The Manchester, Bolton & Bury Canal's main trade was in coal. Open narrow boats were regularly used, but with a difference. Instead of loading the coal directly into the hull, the boats were supplied with eight boxes, each holding approximately 1.5 tons that fitted neatly into the hull, where they were secured. Each box had a metal frame, allowing it to be lifted in and out of the boat with ease. This was far more efficient for loading and unloading than shovelling everything in and out. It took a while for the rest of the shipping world to catch onto the idea of loading a boat with standard units. When they did, they called it containerisation.

Alec Waterson worked at the Ladyshore boatyard on the Manchester, Bolton & Bury, building boats for the local colliery. He has written a very full account, and the work he describes probably changed little over the years and is typical of small yards all around the system. It was not a grand place, just a small office block with a long shed attached, big enough to hold one boat during construction. The side next to the canal was left open, so the boat could be launched sideways down a slip into the canal. The boat was built up on a frame known as the cradle. First the bottom was laid down, made out of 3in-thick elm planks. These were sawn to the correct size and shape using a template to create the outline of the finished vessel. Next the stern and stem posts were carved out of solid pieces of oak using an adze. This tool can be seen in illustrations of ship building as far back as medieval times. Now the side had to be built up, and the top and lowest layers of planks or 'strakes' were made out of oak. There is an interesting problem here. The sides of a vessel have to curve and planks are not usually flexible. But they are if you steam them. The wood was heated in an iron chest, filled with steam from a boiler fixed at one end. While still hot, the

A pair of narrow boats breasted up; the motor on the left, the butty on the right.

plank was bent to the frame and fixed. The space in between the oak strakes was filled with pitch pine planks. The boat was then given a sheath of planking on the inside, covering the frame members, the knees.

The boat had to be made watertight. Nothing was ever done to the elm bottom. The wood absorbs water and swells, closing every joint and not leaving so much as a hair's breadth of a crack. The exterior planks were treated by the time-honoured system of caulking. Oakum was originally old rope teased out, often by convicts. It was laid into the seams between the planks and pushed into the crack using a caulking iron and mallet. The job was completed by covering the seams with pitch. The inside of the boat was treated with an unsavoury mixture of tar and fresh horse manure. When everything was complete, the boat was launched sideways from the shed, down the slipway and into the water. This is a much simplified version of the process, but what is fascinating about Mr Waterson's account is that there is nothing in it that would seem the least bit strange to a medieval boat builder. It is a useful reminder that, though the narrow boat first appeared in the eighteenth century, it was also a part of a craft tradition stretching back for generation after generation.

Everything written about boats and barges so far has described them as working under sail, being towed or quanted. Yet we are talking about the years of the Industrial Revolution when steam power was being applied to more and more different processes. Why was it not seen more on the canals, at least by the end of the eighteenth century? In fact there were a number of experiments with steam on the water, starting in France in the 1770s. The most successful of these attempts was not due to some greasy-handed engineer, but was the work of the Marquis Claude de Jouffroy d'Abbans, whose paddle steamer *Pyroscaphe* puffed up the Saône in 1783. But it was never developed. The 1780s were not a good time for engineering advances in France, and were even worse for aristocrats. Things were very different in the industrial heartland of Scotland, centred on Glasgow and the Clyde.

The story begins at Leadhills in the Lowther Hills, south of Glasgow. This was at the heart of an important lead mining district, which gave the village its name, and it was here that William Symington was born in 1764. He received a good education and his family hoped he would go into the ministry. But he and his brother George had other ideas, and began working on building a steam engine for the nearby mines at Wanlockhead. The young man's initiative so impressed the mine manager that he encouraged Symington to go to Edinburgh University to study engineering. It is not at all clear why he turned away from mining in the hills to steam on the water, but perhaps it was the challenge that appealed. It was generally thought that lighting fires on boats was far too dangerous, so he decided to put it to the test. He put an engine in a pleasure boat on a loch near Dumfries in 1788 and survived to tell the tale, as did one of the passengers, the poet Robert Burns. Encouraged by this successful experiment, he began working seriously on developing practical paddle steamers. He was encouraged by Lord Dundas, a great enthusiast for improving transport in Scotland. The experiments culminated in the launch of the *Charlotte Dundas* in 1803. The mechanism was very simple. Symington designed the engine himself and had it built at the Carron Ironworks near Glasgow. The 22in-diameter cylinder was set horizontally and drove the paddle wheel through a connecting rod attached to a simple crank. On 28 March 1803, it was given a full trial and succeeded in pulling two fully laden 70-ton barges along an 18½-mile stretch of the Forth & Clyde Canal in nine and a half hours.

Charlotte Dundas was the world's first steam tug and the trial seemed to have been a success. The proprietors of the Forth & Clyde, however, had reservations. They were concerned that the wash would damage the banks of the canal and decided to abandon the experiment. Symington was to have had one further chance to prove his vessel, when the Duke of Bridgewater asked for a trial on his canal. Just days before the appointed day, the Duke died and no one else took up the offer. The opportunity to introduce steam onto the broad canals of Britain had gone, and it was to be many years before development restarted.

The Forth & Clyde was, in time, to become home to one of the best-loved of all small cargo steamers, the Clyde Puffer. By the middle of the nineteenth century, steamship design had moved forward – the paddle wheel had largely given way to the screw propeller and engine efficiency had been greatly increased. Having rejected the *Charlotte Dundas* the Forth & Clyde Company decided to have a second attempt at introducing steam to the canal. An engine was installed in the 80-ton lighter *Thomas* and trials took place in 1856. The results were very satisfactory and from this simple beginning the puffer was born, still recognisable as a close relation to the original lighter. The puffers were flat bottomed, bluff in the bow and rounded in the stern and developed specifically to fit the locks of the Forth & Clyde. In later years the puffers moved out of the canal into coastal waters and became the workhorses of the highlands and islands up and down the west coast. The coastal steamers were rather more sophisticated than the canal boats, and these are the vessels described here.

They originally had an open steering position, but later a wheel house was added directly behind the funnel, which makes steering quite interesting. There was a captain's cabin in the stern and crew quarters in the fo'c'sle. The space in between was one big hold. The foredeck had a single mast, steam winch and derrick. This was among the first vessels to be fitted with a compound engine. Steam entered the first, small cylinder under pressure, but there was still pressure left in the exhaust steam, so it was then fed into a second, larger, low-pressure cylinder. After that it was allowed to escape up the funnel, with a puff at each stroke – hence the name puffer. Salt water and boilers is not a good mixture, so the engines had a condenser added. The condensed steam was pure and could be used again, a very practical solution, but it meant that the puffers no longer puffed.

They became famous through the Para Handy Tales of Neil Munro. Para Handy was the notoriously unreliable skipper of *The Vital Spark*, a vessel of which he was inordinately proud:

Oh man! She wass the beauty! She was chust sublime! She should be carryin' nothing but gentry for passengers, or nice genteel luggage for the shooting-lodges, but there they would be spoilin' her and rubbin' all the pent off her with their coals, their sand, and whunstone, and oak bark, and timber, and trash like that.

One puffer, at least, has been released from the duty of carrying coal and other trash, and adapted for carrying passengers, though whether they would all qualify as gentry is another matter. During the Second World War, puffers were built for the important job of supplying the navy with everything from munitions to tins of beans. They were Victualling Inshore Coasters or VICs. *VIC 32* survives much as she was when built in 1943, still powered by steam from a boiler fed with coal. The one difference is that the hold has been adapted to create two levels: an upper level for the saloon and cabins down below. She is not the speediest craft in the world. On one occasion when I was working in the engine room, word came down the speaking tube from the wheelhouse that we were approaching the measured mile on the Scottish west coast, an ideal opportunity to see just how fast she would go. I got the pressure up to the point where the needle on the gauge was hovering around the red mark, above which the safety valve would have blown, and kept it there. We approached the mile going flat out and at the end we had achieved a stately 6 knots, which was what she always did. The old lady was not to be hurried. I have travelled on the *VIC 32* down the Crinan and the Caledonian Canals, though never the Forth & Clyde, her true home waters. But these trips were enough to show what an

1 The two Harecastle tunnels on the Trent & Mersey; the narrow Brindley tunnel is on the left, the broader replacement on the right.

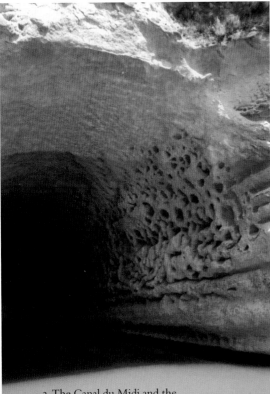

2 The Canal du Midi and the world's first canal tunnel.

3 A traditional narrow boat on the Coventry Canal.

4 Pigeon Lock on the Oxford Canal.

5 The five-lock staircase at Bingley on the Leeds & Liverpool.

6 A rural section of the Huddersfield Narrow Canal near Marsden.

7 The Llangollen Canal has always been one of the most popular with pleasure boaters.

Opposite, clockwise from top left

8 A peaceful scene on the Kennet & Avon near Bath.

9 The elegant Dundas aqueduct on the Kennet & Avon.

10 Telford's first experiment with a cast-iron aqueduct at Longdon-on-Tern.

11 The deep cutting at Laggan on the Caledonian Canal.

12 Rennie's Lune aqueduct on the Lancaster Canal.

13 A busy scene at Fort Augustus on the Caledonian Canal.

14 A hire boat on the beautiful Brecon & Abergavenny Canal.

15 The city scene: Farmer's Bridge locks, Birmingham.

16 Textile mills and industrial
housing developed along the line of
the Leeds & Liverpool at Skipton.

17 The late Fred Schofield at the tiller of his Humber keel *Comrade*, which he regularly worked on the Sheffield & South Yorkshire.

18 *Albion*, the last surviving Norfolk wherry.

19 The National Waterways Museum at Ellesmere Port.

20 A rare sight even then: working narrow boats being admired by the author in 1976.

21 Working boats at Stoke Bruerne on the Grand Union.

22 The steam-powered narrow boat *President*.

Clockwise from left

23 Fellows, Morton & Clayton boat at Gloucester Docks.

24 Sharpness is still a working inland port.

25 Traditionally painted narrow boats moored alongside boats in the blue and yellow livery of British Waterways in Gas Street Basin, Birmingham.

26 Dereliction on the Dearne & Dove Canal.

27 A deep cutting on the mainly derelict Barnsley Canal.

28 The restored Basingstoke Canal at Odiham.

29 The first successful canal restoration scheme – the Stratford Canal. It is noted for its split turnover bridges that allow a towrope to pass through.

30 Functional elegance: the bridge at Barbridge Junction on the Grand Union.

31 A cast-iron bridge on the northern Oxford, proving that using new materials need not lead to ugliness.

32 Even the most functional buildings have their own beauty, often resulting from contrasting textures; details from a warehouse at Gloucester Docks.

33 A simple wooden bridge across a lock on the Rochdale Canal.

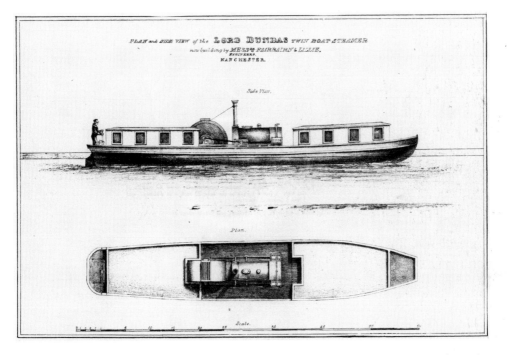

The *Lord Dundas* steamer built for the Forth & Clyde. It has, in effect, a portable steam engine dropped into the middle of the boat, driving a central paddle wheel.

The Clyde Puffer *VIC 32* on the Crinan Canal.

A steam tug hauling narrow boats out of Braunston tunnel; the children wait with the horses, ready to continue their journey.

invaluable vessel she is, every bit as much at home in the confined waters of the canal as out at sea, heading over to Arran or Islay.

Many craft were fitted with steam engines on the broad waterways of northern England and Scotland, but when it came to the world of narrow boats the advantages were not so obvious. Narrow boats have a limited cargo capacity; fit an engine and you have to take valuable space for an engine room and a coal store. The first steam narrow boats appeared on the Grand Junction Canal in the 1860s, but the main steam fleet was owned by Fellows, Morton and Clayton. Between 1889 and 1939 they built up a fleet of thirty-one steamers. Although the steamer had perhaps half the carrying capacity of a horse boat, it was powerful enough to tow another narrow boat behind it. In order to take up as little space as possible, both engine and boiler had to be kept small. The only way to get a lot of work out of a small engine is to use high pressure, and some engines worked at pressures up to 140psi, which is roughly twice the normal working pressure of the *VIC 32*. The Grand Junction was the ideal canal, as the locks were big enough to take both vessels side by side. Even so, the only way the fleet could make a profit was by running a fast service to a strict timetable. This involved working 'fly', travelling day and night, with change crews on board. There would normally be five men on the steamer and three on the towed vessel, the butty. It must have been an interesting experience, as the steerer of the steamer had no control over the engine and the engine driver couldn't see where they were going. They could only communicate through coded signals on a lanyard. It obviously worked, for the typical timetable for a run from Limehouse Basin on the Regent's Canal to Fazeley, Birmingham allowed just fifty-four hours for a journey of 151 miles and 161 locks. The crew who ran these boats were an elite, smartly dressed in white uniforms. The boats themselves were equally distinctive. Known as 'joshers' after Joshua Fellows, they had wooden bottoms, but riveted iron sides and were slender and graceful compared with other narrow boats. One FMC steamer, *President*, is still running, but with a replacement steam unit. Sadly, none of the original FMC engines have survived.

Steam narrow boats remained rare, but steam tugs were more popular, particularly on lock-free canals and broad rivers, where they could tow a whole string of boats behind them. They were also extremely useful in tunnels, greatly speeding the flow of traffic. But the days of steam on the canal were numbered and by the end of the nineteenth century a new and powerful alternative had appeared on the scene. In 1892 Otto Diesel took out an English patent for a new type of internal combustion engine. Over in Sweden another engineer had also been working on similar engines. Erik August Bolinder was born in Stockholm in 1863 and by 1888 he was running the family's engineering firm. His first efforts were concentrated on paraffin engines, but he soon began looking at ways of adapting the diesel engine for marine use. By 1908 he had designed a simple two-stroke engine. All diesel engines develop sufficient heat during the compression stroke to provide ignition, but they need to get started first. In the Bolinder engine, the fuel is first heated in a vaporiser, or hot bulb, using a blow lamp.

The hot bulb engine seemed more than a little alarming, especially when used in the narrow confines of a boat's engine room. The Bolinder salesmen had a spectacular safety demonstration: they used to pour diesel fuel into the hold and chuck in a lighted match – nothing

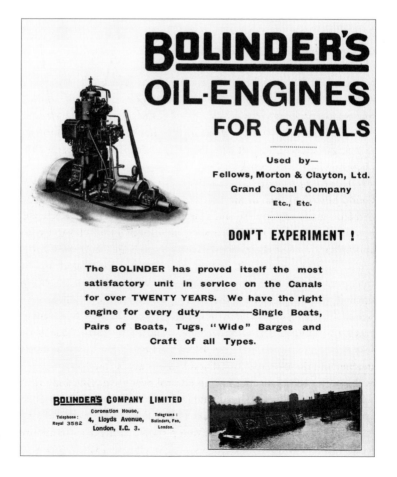

An advert for a Bolinder engine: this was the engine that revolutionised narrow boat transport.

Tom Puddings being loaded at the wharf, while a short train makes its way down the canal.

happened. It was something of a gimmick, but was sufficiently impressive for boat owners to give it a go. In 1910 a Bolinder was fitted to a Thames lighter, and the experiment was a huge success. One advantage was immediately obvious. Once the engine had been started, there was no need to have a driver down in the engine room: everything could be controlled by the steerer. Converting from steam to diesel removed the need for an extra man in the crew. The canal scene was transformed: by 1930 the engines had been fitted to over 200 narrow boats and the distinctive pop-pop of the Bolinder became as familiar as the clip-clop of hooves had once been. On many canals, the single boat was replaced by the working pair. The butty remained the same as the basic horse boat. The motor boat had a different stern, which needed to incorporate the propeller. It was more rounded than the butty stern, with a much smaller rudder controlled by a Z-shaped tiller. The basic motor boat design survives on countless holiday narrow boats today. For the boating families, this was all very good news. One family could run a pair, which meant that they had two cabins to live in, instead of just one. They had just doubled their living space.

The story of canal craft deserves far more than a mere chapter. Even within the comparatively limited category of narrow boats, there were many variations. Boats were built of wood, iron, steel and even concrete. Others were designed for specific cargoes. By the late nineteenth century, for example, the by-products of the gas industry, such as tar, were carried in specially adapted narrow boats with closed holds. Even less romantic were the barges of the Tamar that specialised in carrying manure, though *Shamrock*, the last surviving example, turns out to be an elegant ketch. This rich variety has developed from the need to find vessels suited to particular waterways and different trades. No vessels illustrate this better than the curiously named Tom Puddings.

The Aire & Calder Canal has always been a thriving waterway and scene of innovations. Steam tugs were introduced as early as 1831 and in 1862 William Bartholomew introduced trains of compartment boats to the system. They were known as Tom Puddings simply because the containers were no more than open-topped oblong boxes, like pudding tins. These were loaded at collieries along the canal in trains of up to fifteen vessels, each holding about 35 tons. To make this snake-like train manageable, 'false bows', wedge-shaped boxes, were fastened to the first vessel in the train behind the tug. The end of the journey was at the port of Goole, where special coal hoists were built. The individual Tom Pudding was floated underneath the hoist, picked up, carried to the top of the tower, then upended to discharge down a chute to the waiting collier. Once fully loaded the collier would make the journey down the coast to the final destination.

This system survived into the 1970s, when I first saw it in action, though steam had by then given way to diesel. When you are chugging along in a typical holiday narrow boat it is a little intimidating to see a train of fifteen Tom Puddings heading towards you. I moored up to watch the process of getting them through a lock. On this occasion the train was heading down to Goole, but was far too long to fit into the lock in one go. First the tug went in with seven compartment boats and the lock was operated in the usual way. But that left the rest of the train above the lock with no tug. It would have been possible, of course, for the tug to go back up to collect them, but a far simpler method was used. The lock was refilled, then the top gates were opened and the bottom paddles lifted. The resulting flow of water sucked the rest of the train into the lock, the top gates were closed and soon the train was reunited. It was an impressive sight and a useful reminder that a system introduced over a century earlier could still look surprisingly modern and efficient. If there is one message that has reached me from looking at canal craft it is this. They might look old-fashioned today, but at the time they were designed they were perfectly adapted to the job for which they were built – and that is the only criterion that really counts.

8

THE CARRIERS

THE CANAL carrying companies were as diverse as the goods they carried and the boats they used. Some were large concerns, some small; some covered a whole network of canals while others looked to purely local trade. Even at the very start of the canal age, carriers came into the business via very different routes.

Hugh Henshall was born near Stoke-on-Trent in 1734 and trained as a surveyor. He was taken on by Brindley as an assistant, worked with him on his canal schemes and eventually became his brother-in-law. When Brindley died in 1772, he took over the task of completing the Trent & Mersey Canal. It was a logical step to move from building a canal to running boats on it. He had established excellent links with the likely customers, especially Josiah Wedgwood, the canal's chief promoter, and was assured of a good trade from the start. It proved sufficiently successful for the company to be formally set up as Hugh Henshall & Co. with its own warehouse at Castlefield, the terminus of the Bridgewater Canal in Manchester. As the network spread with the opening of the Coventry and Oxford Canals, he began sending boats as far as London. Wedgwood remained one of his best customers, soon paying over £1,000 a year in freight charges. Henshall was now making a lot of money out of the Staffordshire potters and he could see that they were thriving. He now had enough capital to set up in a pottery business of his own in partnership with Robert Williamson. This was very much a family concern, for Williamson had married Brindley's widow, the former Anne Henshall. The carrying trade seems to have been profitable from the very first.

The arrival of the canals alerted local businessmen to the possibilities of profit. At the eastern end of the Trent & Mersey, a busy little inland port developed at Shardlow, where the canal traffic met the barges of the Trent. James Sutton is first mentioned as 'a village official', but he was quick to spot a likely source of trade and built the Navigation Inn, which still survives. It obviously did well and soon business was expanded and he added a butcher's shop. We actually have a description of what the pub was like, because it was visited by a German tourist, Charles P. Moritz, in 1782, where he met a group of colliers – not miners, but men who worked coal boats:

> A rougher or ruder kind of people I never saw than these colliers, whom I here met assembled in the kitchen, and in whose company I was obliged to spend the evening. Their language, their dress, their manners, were all of them singularly vulgar and disagreeable; and their expression still more so; for they hardly spoke a word without an oath, and thus cursing, quarrelling, drinking, singing, and fighting, they seemed to be pleased, and to enjoy the evening.

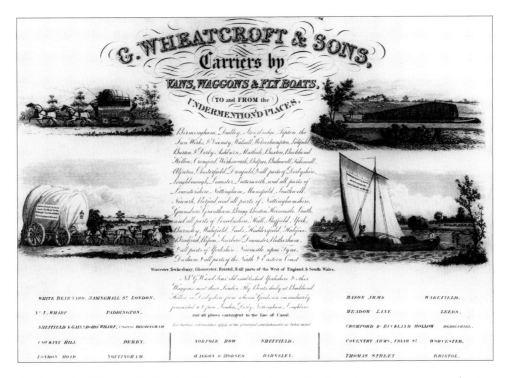

Carriers such as Wheatcroft offered a full range of carrying services from stage wagons to narrow boats.

Having damned them, he then went on to point out, 'I must do the justice to add, that none of them, however, at all molested me. On the contrary, every one again and again drank my health.' With the profits from this boisterous business, the Suttons were able to expand. They acquired a very useful site on a triangle of land, bordered by the river, the canal and the new turnpike road and it was here that they built their warehouse and began a boat building and carrying business. They constructed both narrow boats and sailing barges for the Trent.

The Suttons traded widely over the system, and carried anything. One of their more unusual cargoes was a set of five bells, cast at the Whitechapel Foundry in London, for Shardlow church. Their bulk cargoes were largely malt and coal, but nothing was too small. An inventory for one load featured: '1 box starch, 1 hogshead sugar, 1 chest tea, 1 bag coffee, 1 tin molasses.' Like Henshall, they prospered and an early nineteenth-century letterhead describes the company as 'Carriers on the River Trent and Canals also General Wharfingers', offering a daily fly boat service to Liverpool, Manchester, the salt works of Cheshire and the potteries of Staffordshire.

The other main carrying company in Shardlow was run by the Soresby family. They came to Shardlow specifically because they had heard about the plans for the canal and decided to get in quickly, buying up land on either side of the proposed line. They ran packet boats, vessels that carried both cargo and passengers. And, like the Suttons, had their own boat-building business as well. They traded far and wide and had agents as far away as London in the south-east and Liverpool in the north-west.

These were concerns that prospered. Something of the wealth generated in Shardlow can be seen in the Georgian houses near the canal, and even the old Soresby warehouse, with its semicircular windows, has something of that same air of eighteenth-century elegance. And with their prosperity the town began to grow. When the canal opened the population was no more than 200; by 1841 it was registered as 1,306. This was the time when the canal carrying trade was at its height. Railway competition was to change all that. In the next half century the population halved and the names Sutton and Soresby disappeared from the list of carriers.

One name that is still well known in the carrying trade is Pickford's, though today we think of them mainly when it comes to moving house and as road hauliers. Indeed, it was as a road haulage firm that they started out in business. They began at Poynton near Manchester and by 1756 they were advertising in the *Manchester Mercury* as James Pickford, the London and Manchester Waggoner, advising customers that they had moved their London business from Blossom's Inn to the Bell Inn. They offered a twice-weekly service in each direction. This sounds somewhat insignificant, just four wagons a week, but it in fact represents a very considerable undertaking. These wagons were great lumbering vehicles that took a team of eight horses, with a ninth taken along in case of accidents, and the journey could not be done in less than a week – and in bad weather could take even longer. To keep the service running Pickford's would have needed a minimum of eight wagons and over seventy horses; this was a considerable investment.

James Pickford died in 1768 and his widow took over the running of the business until her son, Matthew, was old enough and experienced enough to take charge in 1772. No one operating out of Manchester at that time could have been unaware of canal developments and the way in which the network was expanding. Pickford's were wary. The new system could well take away their trade, so they began buying boats as an insurance policy and operated a limited service. It was a modest effort. In the 1780s they had twenty-eight wagons and only ten boats, but they must have been very aware that of the 250 horses they owned, only ten were needed for the boats and the rest were pulling the wagons. Horses are expensive beasts to run: economic arguments seemed to be pointing in one direction. But still Pickford's waited; there was still no direct link by water between Manchester and London. Work on the Oxford Canal had ground to a halt in the economic gloom of the period, but by the 1780s confidence was rising and it looked as if work on the Oxford would soon be restarting. Once it was completed, the through route would be open, offering direct competition to the wagon service. Now was the time to act, and they began negotiations. The Canal Company was still short of cash, so Pickford's were able to get favourable terms for a lease for a warehouse and wharf at Banbury. Now they needed more boats and found just what they wanted; they bought out Hugh Henshall, and as well as acquiring boats they were able to take over his old warehouse at Castlefield in Manchester.

The new route was not exactly straightforward: down the Bridgewater to the Trent & Mersey, on via the Staffs & Worcester to Birmingham, down the Coventry to the Oxford, eventually arriving at the Thames. It was far longer than the road route, but there was a huge saving in manpower and, more importantly, in horse power. Even so, Pickford's were not yet fully committed to canals as being better than their old road business. The promotion of the Grand Junction Canal in 1793 changed everything. This offered a far superior route and even before it had opened, Pickford's had established a base at Paddington. Now they began to concentrate on canal business and were soon expanding their fleet of vessels and offering a fly-boat service between London and Manchester. In the event, they proved over ambitious and suffered the familiar problem of rapid expansion – cash flow difficulties. The money being paid in freight charges was failing to catch up with the money owed to the canal companies in tolls. The family do not come out of the story very well: two of the family sold their shares before the debts were due to be settled, leaving two others to try and keep the business afloat, literally and metaphorically. They failed and although Pickford's continued in business, it was no longer with members of the family at its head. But once the crisis had passed, the company prospered again. With the opening of the Regent's Canal in 1820, they acquired a new London base at City Road Basin. An account of the workings of the concern at the peak of its prosperity was published in *The Penny Magazine* in 1842. The picture it paints is of a vast, very efficient organisation.

City Road Basin consisted of a general warehouse, simply described as 'huge' and two other warehouses. The shipping warehouse was for goods being sent out of London and the discharging warehouse for goods arriving on the boats. Each had its own wharf where boats could be drawn up alongside. There were extensive stables and offices and a counting house, employing over a hundred clerks. The basin was at its busiest in the early morning, with wagons arriving from all over London to collect goods and late afternoon when there was another onslaught of wagons providing goods for despatch. The evening was the most frantic time, as Pickford's prided themselves on getting goods away as rapidly as possible. Everything had to be recorded, weighed, ticketed and stowed. By late evening goods that had arrived on carts were on their way in half a dozen boats. It all sounds tremendously efficient, and no doubt it was, but the methods of deciding tariffs seems to have been more than a little arbitrary. With such a variety of goods being sent prices were often made up on the spot, and clerks had a keen eye for what the sender might be able to afford. 'Nobility must expect to pay for their titles', as one put it.

Pickford's ran stage boats, which were comparatively slow, and fly boats, mainly for perishable goods for which a premium had to be paid. Records show that the fly boats were indeed rapid in canal terms averaging 2mph day and night, which is admittedly only a modest walking pace, but a vessel on the Grand Junction, for example, going from Bull's Bridge to Fenny Stratford in a day would not just have the 48 miles to cover, but would have to pass through sixty-six locks as well. This seems impossible to anyone who has ever travelled the canals, but the fly boats had the law on their side. Canal rules laid down that they had priority over other craft and could claim first place at every lock. There is, of course, a difference between making a rule and enforcing it. One cannot quite imagine a boatman, with a schedule of his own to keep, cheerfully tipping his hat to a fly boat skipper and politely ushering him forward. One of the more difficult jobs of the lock keeper was ensuring the priorities were preserved. The problem only vanished when the stage boats moored up for the night and the fly boats had the waterways to themselves. At the time when the description of City Basin was written, Pickford's had a fleet of eighty boats and 400 canal horses.

The Penny Magazine journalist described a thriving, prosperous business that looked set to continue into the indefinite future. But Pickford's' reputation for keeping an eye on the broader picture was well earned. The nineteenth-century owners were as canny as their eighteenth-century predecessors. The Grand Junction Canal had a new neighbour following a very similar route. The London & Birmingham Railway opened for business in 1837 and work was then also under way on the Manchester & Birmingham Railway. That meant that there would soon be a faster route between Pickford's two main bases in London and Manchester. Just as they had been willing to drop their road haulage business once the canal network began to spread, now they were equally quick to abandon boating. The fleet was steadily reduced and in 1847 they announced that they would leave the canal business altogether. Rail transport was very different from either road or canal in that money was not made by charging tolls for their use. Instead the railway companies ran their own trains. Pickford's could not break into the new market, but returned to where they had begun a century ago, back on the roads.

Pickford's were by no means the only company who recognised a powerful and dangerous competitor when they saw one. The Grand Junction Canal Company ran their own boats and relied on those as much as they did on tolls from other carriers to ensure a profit. Not surprisingly they did not greet the arrival of Robert Stephenson and his proposed railway line from

Birmingham to London with any great enthusiasm. In 1834 Stephenson had set his men to work on the construction of a long embankment across the Ouse valley and he planned to use the spoil from a nearby cutting to build it. But to get his carts from one to the other he had to cross the canal. He applied for permission to build a temporary bridge, and was given a flat refusal. Stephenson bided his time, but on 23 December, when canal company officials were more concerned with getting ready for Christmas than they were with railway builders, Stephenson gathered a large group of navvies together. They advanced on the canal at Wolverton at night, and set out building a bridge by torchlight. When the canal company officials got up the next morning the bridge was built. That was not the end of the story. What one band of hefty workmen could build, another could demolish – and did. In the end the affair was settled in the courts, and the railway won. The canal companies could no more stop the advance of railways than the stage coach and road hauliers had been able to prevent the arrival of the canals. Eventually the Grand Junction Canal Company followed Pickford's lead and abandoned carrying on their own canal. But it certainly did not mark the end of canal traffic, as new companies moved in to fill the void. The most famous of them all was an organisation that we have already met, Fellows, Morton & Clayton.

Fellows was the first to start up in business. James Fellows began carrying on the Birmingham system in 1837 and when he died his son Joshua inherited a thriving concern. He did not share the pessimism of Pickford's or the Grand Junction carrying company and his optimism was rewarded. When the Grand Junction stopped carrying in 1876, Fellows bought up many of the boats and formed a new company, the London & Staffordshire Carrying Company. All this needed extra capital and he was joined by businessman Frederick Morton to help run the growing concern, which now became Fellows, Morton. Two thirds of the partnership was now in place.

There was another successful carrying company based near Birmingham, begun by William Clayton, that went into business in 1842. By 1862 he was doing well enough to move to Saltley on the Birmingham & Warwick, which, as well as being a base, was developed as an important boatyard. It was about this time that Clayton's realised there was a good business to be had moving tar around the system. It was a decision that seems so logical that one wonders why nobody else thought of it. The gas industry was hugely important by the middle of the nineteenth century, when gas lighting was no longer a luxury, but considered an essential for any self-respecting town, and tar was a valuable by-product. The obvious place to build a gas works was next to a canal, where they could have the coal delivered to their doorsteps. Clayton's began building their tanker boats and established a virtual monopoly. Ironically, one of their best customers was the old rival, the railways. The Great Western Railway established a sleeper depot at Bull's Bridge and they needed a regular supply of creosote for treating the timber.

The new enterprise was so successful that Clayton decided to abandon general carrying altogether, and sold off that part of the business to Fellows and Morton in 1889 and Fellows, Morton & Clayton was born, generally known by the less cumbersome initials, FMC. Why were FMC so successful, when other companies were abandoning the whole idea of canal carrying? We have already seen that they were innovative in introducing steamers. In fact, Clayton's also had three steamers in service by 1888, before the merger. But the real secret was in organisation. Steamers may have been glamorous, but by the end of the century they only represented twenty vessels out of a fleet of over 200. That is a lot of boats, and in order to make a profit they had to ensure that there was always a boat available where a cargo was waiting to be picked up and that boats spent as little time as possible running empty. That could only happen if there was an efficient organisation, tracking boat movements. Boatmen knew the waters they travelled regularly, but an organisation like FMC needed to know everything there was to know about the whole system. They did so, thanks to their chairman, Henry Rodolph de Salis.

Fellows, Morton & Clayton were probably the most famous carriers on the canals and the most innovative. This is their steamer *Baron*.

He was an unlikely figure for the head of a canal company. His father was an aristocrat and a noted diplomat, with extensive estates in Ireland – Count John Francis William de Salis. Henry, the younger son, was educated at Eton, not an institution noted for turning our engineers. He was obviously the odd one out in the family. He joined the company before the First World War and made it his personal task to explore every single yard of the whole system, not ignoring even the shortest branch. He had his own private launch and in eleven years he travelled 14,000 miles of rivers and canals. The result appeared in *A Handbook of Inland Navigation*, the Bradshaw of the waterways. One of the remarkable features of this publication is the number of river navigations that were once used for trade, with names that have now been all but forgotten, such as the Ipswich and Stowmarket Navigation. My edition comes with the terse note: 'The navigation is very narrow and tortuous. Not much trade is done above the Patent Manure Works at Bramford.' Nevertheless, it was recorded as navigable for nearly 16 miles. De Salis also gave the canal world a useful name for all those people who are quite happy just watching boats go through locks or glide under bridges – 'gongoozlers'. His definition reads 'an idle and inquisitive person who stands staring for prolonged periods at anything out of the common'. It is a splendid word that ought to be in everyone's vocabulary.

The main value of the work lies in the wealth of detail it supplies. Distances for every single wharf along a canal are listed in miles and furlongs, usually with reference points at half-mile intervals. So anyone setting out from Birmingham for, say, Arkwright's Wharf would know that they had 19 miles 4 furlongs to travel, with thirteen locks and a 433-yard tunnel to pass through. Given this sort of detailed information, the company could work out accurate timetables, which became much more important after 1912. That was the year they decided to try out a Bolinder engine for the first time, fitting a modest 15hp model into the narrow boat *Linda*.

They decided the experiment was a success, and began building motor boats and replacing their old fleet of steamers. As soon as the motor boats were well established, they set up a system known as 'ockers'.

Ockers consisted of working pairs, where the motor boat and the butty each had its own captain. The butty crew had the job of working the locks. Because these were only temporary partnerships, motor boat and butty did not have to be heading for the same final destinations. It might be, for example, that a pair left London and the motor was carrying a load for Leicester and the butty was bound for Birmingham. The butty would be dropped off at Norton Junction and then picked up by another motor that happened to be available for the rest of the journey. Alternatively, the butty could stop off at one of the many horse stations managed by the company and continue as a single horse boat. It all made for great flexibility, but it depended on the company knowing just where everyone was every day – which was not necessarily where the plans said they ought to be. So every captain was given a bundle of stamped addressed postcards, which he popped into a letter box at the end of the day with his position marked. Today we have all the advantages of computerisation, but the mail actually takes longer than it did a century ago. With luck you will get a first class letter the next day. In those days a halfpenny stamp was all that was needed and a card posted in London in the morning would be at the Birmingham office the same day.

The advantages for the company were obvious, but flexibility extended much further with the workforce. The company decided who did what, where and when. A man might be taken off a boat and set to work at the wharf, loading and unloading, while he was waiting for a place to dock. That could mean extra money in some circumstances, but there was a niggling resentment at having so many controls imposed by management. It worked well as long as trade was booming, but in the 1920s the situation began to change very dramatically. At the end of the First World War a very large number of former army trucks were suddenly available at very low prices. Rail traffic was no longer the threat: now it was the rapidly growing road transport system that was eating into canal trade. The trucks had a huge advantage over both canal and rail – they could offer a door to door service.

FMC had a valuable trade in sugar that was being directly threatened and the only way they could compete was by cutting prices; the easiest way of doing that was to cut wages. For a captain this could mean a considerable loss in earnings and to make things worse all bonuses were cancelled. The boatmen called a strike, which received the full backing of the Transport and General Workers Union. FMC refused to recognise the Union's right to get involved and refused to negotiate with anyone. The captains brought their boats to Braunston, and chained the whole lot up, so that nothing could move. The company sent down a truck under police escort with orders to cut the boats free. The men stood firm, there was a scuffle and the outnumbered company men retreated. The police, probably wisely, decided this was a private battle and refused to get involved. The company responded by sending in a private police force to break the strike and the violence became altogether more vicious with even the boatwomen being attacked. The strike lasted for seventeen weeks and went down in folk lore as 'the battle of the canals'. Eventually negotiations did take place: the Union was recognised and the pay cuts were reduced. The men claimed a victory, but in truth the only real winner was out on the road in a truck.

FMC was not the only partnership involving Joshua Fellows. He was a man always on the look out for ways of expanding his watery empire, and he was to challenge one of the longest established carrying firms on the canals. The Danks family had a long history of trading on the River Severn, and in 1804 they added narrow boats to their fleet of trows. Stourport was their headquarters, an ideal location, where river vessel and canal boat could exchange cargoes.

The Severn & Canal ran everything from coasters to narrow boats.

Eventually they expanded and a new company was formed by amalgamation, Danks & Sanders. The river was ideal for the introduction of steam power, and soon they were running both tugs and small coastal steamers. Fellows interrupted this happy progress by going into direct competition on the Severn. It was clear that there was not enough trade for both concerns to prosper, so they took the very sensible decision to merge in 1873, forming The Severn & Canal Carrying, Shipping and Steam Towing Company Limited, trading up into the English Midlands and down the Bristol Channel to South Wales. Eventually the name was shortened to the more manageable Severn & Canal Carrying Company Ltd (SCC).

One of the most remarkable sights on the river was a steam tug towing up to twenty narrow boats, in two parallel strings. It must have been very strange for a captain at the end

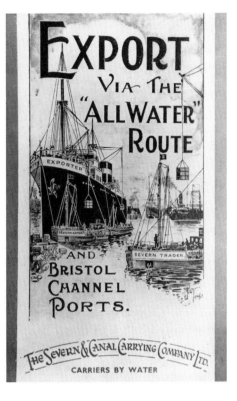

of the line, with the tug as much as 200 yards away. In time the coasters were to make ever longer voyages, often crossing the Irish Sea. There was no other company with quite such a variety of craft under their control and by the 1930s they had their own fleet of lorries as well, offering a collection and delivery service in the Birmingham area. They were the unquestioned leaders of the river trade, but one other company also came to importance, though not strictly speaking as a competitor.

In 1879 Richard and George Cadbury opened a chocolate factory at Bournville beside the Worcester & Birmingham Canal. At the time this was open country, outside the Birmingham urban area, and it was here that the Cadbury family built a model village for the workforce. Industrial villages were not new. Saltaire by the Leeds & Liverpool is a splendid example, with its church, almshouses and other public buildings. But the housing is as regimented there as in any other industrial town of the period. Bournville was to be very different. The roads were not limited to straight lines, houses had gardens that the Cadbury family planted with fruit trees and there were large open spaces. The family were also great believers in physical fitness and provided a swimming pool and sports centre. It was the forerunner of the garden cities on one hand and leafy suburbia on the other. The siting of the factory itself was key to the whole development. It was decided to rely on water transport, so it had its own wharfs and was to run its own fleet. They were innovative in creating a community at Bournville – even the works was known as 'the factory in a garden' – and they were no less pioneering when it came to transport. FMC may have tried their first motor boat in 1912, but Cadbury beat them to it. In 1910 they started trials with motors and when they were happy they had narrow boats specially designed for use with Bolinders. The first was launched in 1911 and must have created a sensation at the time, for it was unlike any other narrow boat then built. It was steel hulled with a sharp bow and a built up hold, covered with hatches that ran straight down to the back cabin. Even the latter was untypical, having little windows and folding beds at either side. The boat

Ovaltine was one of many companies that ran their own fleet of boats.

was steered by wheel instead of the conventional tiller. It was simply called Bournville and over the years the fleet was extended by yet more Bournvilles. At first they were classically minded and added I, II and III, but then started all over again with Bournvilles 1 to 21. The vessels were always immaculate and the lettering was specially done at the works, and was familiar to chocolate lovers everywhere.

The two main cargoes supplied by water were chocolate crumb – a dried, crystalline mixture of sugar, milk and cocoa solids – and evaporated milk. These were the basic ingredients of milk chocolate. There were a number of sites around the country for providing the milk. The first was established at Knighton on the Shropshire Union, but one of the most important depots dealing with both crumb and milk was at Frampton on Severn on the Gloucester & Sharpness. Later, more conventional narrow boats were used and in later days one particular motor boat *Mendip* worked full time between Knighton and Bournville, making two round trips each week. The captain, Charlie Atkins, was inevitably known as Chocolate Charlie. Cadbury took their last boat delivery in 1968.

Not all the cargoes were on Cadbury's own boats. Tom Mayo recalled the hard work of the run. In the 1930s he would be taking twenty hours to get from Gloucester to Bournville. He would leave at 4 a.m. and be at Tewkesbury by 6 and Worcester at 10. He would be out of the top of the Tardebigge flight at 9 p.m. and reach the factory at midnight. This is a journey of over 50 miles which includes working the longest flight of locks in Britain. There were compensations, even if they were unofficial. He helped himself to chocolate crumb, and sometimes to a tin or two of milk – the space in the hold was filled by a brick. And on the return journey when pressure was off, he would stop by Stoke Works on the Worcester & Birmingham and fish for conger eels, which enjoyed the salty outlet from the works. There was another advantage of a more leisurely trip for a young, single man: he could chat up girls walking along the towpath and offer them a boat trip.

When British Waterways began carrying they removed the familiar roses and castles from their fleet, but the boating families still managed to add decorative ropework to the sternpost of the butty.

There were many other carrying companies, some of which were specialised, such as Samuel Barlow, famous in the coal trade, and others were general carriers, like the Grand Union Canal Carrying Company (GUCCC). There were fleets run by individual companies, including a close relation of Cadbury – Ovaltine. The Ovaltine boats also carried the recognisable trade lettering, though curiously it should never have been like that. The malted milk drink was invented in Switzerland and marketed as Ovomaltine, but when the trade name was registered in Britain it was misspelled and no one ever got round to changing it. But in among these fleets, large and small, were the owner-boatmen, the Number Ones. There were never very many of them, perhaps no more than a couple of hundred boats at the most, and mostly operating on the Midland canal system. They normally owned either a single boat or a pair, usually horse drawn. One exception was Joe Skinner, who worked in the Oxford Canal, and who always preferred working with a mule. I never did discover why. The Skinners were typical of the Number Ones, strongly independent, hard working and a family whose whole life was centred on the canal and the canal community. After Joe's death, his widow, Rose, had a home 'on the land' as it was known, but she was never really happy in a house. She would go down to Hawkesbury Junction to spend the day in their old narrow boat *Friendship*, where she was always ready to offer a cup of tea and chat over the old days and bring out photos of Joe with various mules. The Skinners have gone, but *Friendship* survived, if a little worse for wear, and now has a place of honour in the Ellesmere Port Boat Museum.

It is all too easy to romanticise the life of the Number Ones, but independence comes at a price. If there are no cargoes there is no pay; if the canal freezes there is no income. The aim of most was to get some sort of regular contract. Joe Skinner carried for the Morris Radiator factory at its canalside site in Oxford for about twenty years and after that he moved on to carrying for a dairy in Banbury. Even then the loads were one-way and it was always a lottery

The days when companies ran horse-drawn barges on the Dearne & Dove are now only remembered in pub signs.

as to whether there was a load available for the return journey or not. And without a back load that journey went unpaid.

In 1947 a brand new carrier appeared on the scene. Where most carriers started with one or two boats, and gradually built up on that, this one went into business with a ready-made fleet of roughly 1,300 vessels. It was the British Waterways division of the British Transport Commission. Nationalisation had put the canal network under their control, and though individual carriers remained independent, the fleets belonging to the canal companies themselves passed straight into public ownership. About 500 of them were the maintenance vessels needed to keep the canals open and the largest carrying group consisted of the barges and compartment boats on the Aire & Calder. The biggest narrow boat fleet that was taken over was that of the GUCCC, with its 126 motor boats and 130 butties. Other companies decided it was time to leave the waterways, now that there was a buyer in the market. FMC had never really recovered from the 1920s strike, so they sold out and British Waterways acquired another hundred boats. The two fleets were now united to make a grand total of 368 boats, the largest narrow boat fleet the waterways had ever seen. There was an obvious problem: neither GUCCC nor FMC was making a profit and there was no reason to suppose the new organisation would do any better. The pessimists were justified in thinking nothing would really change. The losses mounted and the final blow came in the severe winter of 1962–63, when the canals froze and not a boat moved from the end of December for the next thirteen weeks. By the time the canals reopened, many of the few remaining customers had made alternative arrangements. The great days of the narrow canals as major freight routes were coming to an end.

In 1963 the British Transport Commission was abolished, and the canals passed to a new authority, the British Waterways Board. One of their first actions was to announce the end of narrow boat carrying. One cannot blame them. Travel on the Grand Union today and for quite long stretches you can look across and see the trucks trundling along the M1. The narrow canals could not compete.

Trading in freight on the narrow canals has virtually finished, though the occasional boat can still be seen, often delivering coal in sacks to waterways sites. With its end, a whole way of life has also disappeared from the canals. It is unlikely ever to return, but it is as well to remember that the boating community of Britain served the country and its industries well for some 200 years.

9

THE BOATING COMMUNITY

WHEN THE first canals were being built, the men who worked on them were mainly local labourers. Once they were complete it seems that it was also local men who were recruited to run the boats. The interesting question is: what were they doing before they settled for a life on the waterways? There are many different answers. Folk lore has it that large numbers of gypsies were employed, largely it seems because the traditional painting on narrow boats reminded people of the decoration of gypsy caravans. The trouble with that argument is that the early canal boats were never decorated in this way. Harry Hanson, who was among the first to do extensive research on the boatmen, surveyed the known names of boatmen and found that at best there was only a small percentage that had what were generally regarded as gypsy names. Even then, quite a lot of those names were common in the general population, particularly when you discover that Jones, for example is listed as one of those names, and that the Jones names mostly turn up either in Wales or in the bordering counties. Hanson's conclusion is that gypsies made up no more than 10 per cent of the entire boating population, and could well have been considerably fewer. In his book *The Water Gipsies*, A.P. Herbert has his hero, the canal boatman Fred, explain that the narrow boat decoration comes from gypsy boaters, but I have never heard any actual boatman claim the same, and the name 'water gypsy' is generally resented. On balance, the evidence for a large gypsy population on the canals is at best flimsy, and the evidence against at least as strong.

So if gypsies are out, who takes their place? Another popular notion is that the first boats were worked by former sailors. Again, there is no evidence to support the idea. Certainly the sailing barges in the coastal trade would have had exactly the same crews when they moved onto the river navigations and broad canals, but that is about the best that can be said. To me, it seems unconvincing. The skill of the narrow boat captain has little resemblance to that of the deep-water sailor, beyond being able to handle the tiller. Other skills, including the all-important one of being able to work with horses, have nothing to do with the sea.

Harry Hanson inclines to the view that a lot of the boatmen were recruited from the ranks of carters and small tenant farmers. This seems to make a lot more sense. The world of agriculture was changing rapidly in the eighteenth century. The process of enclosing the waste lands and common fields had begun in Tudor times, but was still hugely important in the eighteenth century. During that time 2,032 Enclosure Acts were passed, involving an estimated area of almost three million acres. The cottagers who had previously had their own strips of land and access and grazing rights to the commons found themselves forced to become labourers

Boating was very much a family business: different generations often working together.

for hire. Their independence was gone, and was bitterly resented. There was a well-known verse of the time:

> The fault is great in man or woman
> Who steals a goose from off a common.
> But who can plead that man's excuse
> Who steals the common from the goose?

In eighteenth-century documents, the cottagers started to appear in the same category as paupers, and the two were often indistinguishable. Anything that offered a regular, comparatively well-paid alternative must have been welcome. Hanson suggests that the tenant farmers supplemented their income with part-time boating. But the arrival of enclosures meant that, in general, the size of farms was increasing and land owners were looking for reliable men who would spend all their time on improvements. It is unlikely that there would have been any time left for boating. The best answer would seem to be a rather disappointingly mundane one. Nobody really knows where the first boatmen came from, but they probably came from many different sources. It was the navvy story all over again; there would always be men willing to work hard at a new way of life if it offered them a better living. Was it a better living? I can do no better than quote a contemporary source on the benefits of enclosure. The labourers would be forced to 'work every day of the year' and 'their children would be put out to work early'. Best of all, 'the subordination of the lower ranks of society' would be ensured. These are not the rantings of some reactionary theorist: the words are taken from an official report from the Board of Agriculture. Who would not think that almost any alternative was worth considering? Many of these men who took to the boats would have employed their own sons to help with the work, leading the horse and working the locks, and so a family tradition would have formed.

One thing that we do know about the early days is that it was rare for anyone to live on the boats. At first, when canals were often self-contained, journeys would have been comparatively short. As the system spread, however, that began to change. In the middle of the nineteenth century when George Borrow was travelling through Wales, he met a boatman based at Llangollen. There are two interesting things. Firstly he was a native Welshman, who declared that although he had passed through many towns he had never met any to compare with Llangollen. This does reinforce the idea that a Welsh carrying company recruited local people. The other is that he described a regular trip, taking slate all the way to London, a journey of three weeks. But even long journeys are not sufficient to account for the movement of whole families onto the boats; after all carters and waggoners had been doing the same sort of trips for generations, and never thought of taking their families with them. It seems that for decades boatmen ran their lives in the same way as the men of the roads, living conventionally in houses and cottages.

The first boats seldom had cabins: they would have been very like the joeys of the Birmingham canals. An early illustration of a boat on the Grand Junction has what looks like a small shed in the stern, with a narrow door, a chimney poking through the roof, but no windows. It was at best a very crude shelter, and certainly not a place to make a home. It seems from a number of accounts that boats were not built with integral cabins, but a carpenter would knock up a simple structure that did little more than keep out the rain. This would have been useful when valuable cargoes were being carried, and the boatman was required to double up as night watchman. There are, however, stories of boatmen sleeping out in open boats as well, where the only comfort was a brazier in the hold, in front of which the boatman would stretch out, toasting his boots.

The big change came with railway competition. The only way the canal companies could compete was by offering lower prices and those inevitably came in part from paying lower wages. For families to make a decent living, there was only one alternative: make the boat your home. This had a double advantage: the family no longer had to pay for a house and they became a working unit, keeping the whole of the crew pay to themselves. It seems to be around this time that the boats became decorated. It is not difficult to understand how this has happened. You have been living a perfectly conventional life in a house with a number of rooms and suddenly everything is squashed into one tiny space. I think anyone in that situation would want to brighten things up to make everything look as cheerful as possible. But it is from this desire for colour that much of the gypsy myth originated. E. Temple Thurston was struck with this when he went to take over the *Flower of Gloster*.

> The exterior of the cabin in the aft part of the boat is gaily painted with vivid reds, glaring yellows, greens, and black. Water-cans, buckets, the shallow pails for the horse's provender, even the horse's traces too – a set of wooden beads strung loose upon a cord – all are painted with their Spanish joy of colour. And this is not the closest relation of their minds to that country which has fed the world with gypsies. I puzzled for a long while before I realised that the scenes which are painted on the panels on their cabin exteriors are rough pictures of those castles, not which you build, but which you find in Spain. What is more, these people are lovers of brass – a sure sign of the gypsy. Inside many a cabin of the boats which ply their long journeys and are the only homes of those who work them, you will find the old brass candlesticks, brass pots and pans, all brilliantly polished, glittering in the light. A brass lamp hangs from the bulkhead.

It is a good description of narrow boat decoration and the furnishings of a cabin, and may have been influenced by the Roma caravan or vardo. But the styles of decoration are actually quite different. The Roma vehicles were generally carved as well as painted and the decoration was largely symbolic with lots of representations of exotic animals from lions to griffins. The boats are famous for roses and castles. You cannot grow roses round your back cabin door, but you can paint them, and your home may not be a castle but you can have your very own idealised version travelling with you. In spite of what Thurston may have thought, if the castles resemble anything they are like those of the Rhine rather than anything you will see in Spain. Brass was not just popular among gypsies in the nineteenth century, and it is an eminently practical material to use on a boat: cheap, cheerful and hard wearing. Whenever I see a genuine narrow boat with traditional decoration, what I see is an affirmation. We may be part of a community that lives apart from the rest of you, but we have as much pride in our homes as any resident of comfortable suburbia.

As more and more families moved onto the boats, they became increasingly isolated from the rest of the community. It was inevitable. They scarcely stopped anywhere long enough to make connections, and when they did stop they used the stores and pubs with which they were familiar and which welcomed them. The Greyhound at Hawkesbury Junction – Sutton Stop to the boatmen –was one of the most popular stopping places, not least because the shop attached to the pub used to let the families have goods on tick. Constantly on the move, it was even difficult to keep contact with members of one's own family. A writer, one of the few who travelled the canals for pleasure in the early twentieth century, described a canal meeting. 'A strange life this barge work! Here today and gone tomorrow. A barge passed us. "How d'ye do, Jim?" "Where for?" "Macclesfield with bricks. What's your game?" "Pleasure trip." No time for more. That was my brother. Haven't seen him a many year.'

It is the fate of outsiders to arouse suspicion, which itself acts to increase the sense of isolation. A boatman, looking back on his working life, said that they were called gypsies, bargees – 'everything bar a gentleman'. Prejudice feeds on itself and it is a sad fact that in general attitudes towards the boaters only really changed when boating was coming to an end. As one old boatman put it, in the working days 'we was bums; now, what are we now, we're marvellous, we're part of the history'. The reality of life was described by a working boatman on the Regent's Canal in the 1880s, who was as critical of the canal company itself as he was of those he met along the way:

> They never tells yer w'en they're goin' to lower the water, nor nothin'. It's never clean, an' it's allers low water, and there's nothin' but naked men a bathin' and thieves wot robs your barge and takes all they can git out of 'er, and blackguard boys wot calls yer names and spits on ye, and throws stones at yer.

Until the introduction of motor boats, and for many families even after that, the most important member of the workforce was the horse. The boatmen generally preferred to break in their own horses, buying them at recognised trading centres, notably Stoke-on-Trent. They also bought regularly from gypsies, with whom, on the whole, they enjoyed a very friendly relationship. In later years, they got horses as and where they could. 'Caggy' Stevens, a famous character on the BCN in recent years had a horse that had started its working life pulling a milk float. 'Bonnie', another boat horse, had an even more unlikely early career, performing in a circus, and even in her working days could still perform her old tricks. Not every horse took to the work. As Rose Skinner, who spent her whole working life on horse boats put it, 'Get a good horse, and there's no further trouble. But a bad 'un leaves you fretted and worn at a day's end.' Boatmen paid for fodder and care of the horse themselves, and if money was short it was the family that went hungry, not the horse. Many people remembered a childhood in which they had nothing to eat all day or perhaps a slice of bread and dripping, while the horse munched away happily. Generally the animals were respected – up to a point. 'I fed 'im, I worked 'im and I belted 'im' was the story for the mule 'Smoker', with a final verdict – 'He was a Christian.' Children often had a special relationship with the animals, though outsiders frequently wrote harshly on the subject. This is Arnold Bennett, describing a scene on the Trent & Mersey in his novel *Clayhanger*:

> Thirty yards in front of each boat an unhappy skeleton of a horse floundered its best in the quagmire. The honest endeavour of one of the animals received a frequent tonic from a bare-legged girl of seven who heartily curled a whip about its crooked large-jointed legs. The ragged and filthy child danced in the rich mud round the horse's flanks with the simple joy of one who has been rewarded for good behaviour by the unrestricted use of the whip for the first time.

Is this a fair picture? We have no means of going back to the nineteenth century to check it out, but Temple Thurston described his boatman as using the whip – not hitting the animal, but cracking it as a reminder of the job in hand, when there was a temptation to stop for a tasty hedgerow nibble. Others looking back on their childhood days remembered their horses with affection. One of the jobs was taking the horse over a hill, while the parents legged through the tunnel. Some took advantage of this rare opportunity to be alone and have freedom for a bit of fun. One boy used to 'play cowboys', galloping the poor old horse, which normally never moved above walking pace, all over the hillside. Sometimes he had to pretend that the horse had caught its hoof in a pothole, when it limped back to resume towing duties.

The horse was almost one of the family; this animal at Banbury on the Oxford Canal in the 1930s is not only enjoying a feed but has a bonnet as protection from the sun.

Children waiting at Chirk in the early twentieth century. The tramway on the wharf brought slate from Glyn Ceiriog.

Nell Cartwright's experience was not so boisterous. Her family were unusual in that they came late to the canals. Her mother suffered from ill health and was told that she needed an outdoor life. She took to boating, with no experience whatsoever. They picked up a first load in Birmingham – HP Sauce and tins of corned beef – and set off for London. Her elder brother was in the fore-cabin; parents, Nell aged eight, two younger twins and a baby in the back cabin. Nell took to the horse straight away, and before the first day was over, she was riding it bareback. The only time she got nervous was when she had to set off on her own with the horse, while the boat went through the first tunnel. She was cold and frightened, but the horse kept whinnying and nuzzling her. 'He's telling me not to be afraid,' she remembered thinking. 'When I got to the end of the tunnel I was as brave as brave.' She never lost her love of the horses:

> I liked my horses and I liked the boat as it was. When they brought the noisy motors out I didn't like that at all. They frightened me because I don't think it was so peaceful and there was no beauty. I mean you look along the boat as it was going and you see that horse just walking along the road and the hedges and trees and everything going by. No one could ask for better than that. What I would like to see now – before I leave this world – I would like to see all the horses come back and the place come back as it was as I knew it.

Nell Cartwright was never to see that dream become reality.

Not everything in Nell's life was that idyllic. Her father was rather more interested in the local pub, where he sang for money, than he was in the hard work of boating. And when her mother had to go into hospital, the children more or less ran the boats themselves. They even legged the

A boating family, with the husband on the motor and wife and children on the butty.

tunnels when necessary. Eventually Nell worked on other boats, but got no salary, just her keep, until she was eighteen. Even then she got less than boys of the same age. She remembered going from London to Birmingham and back in less than a fortnight, during which she never took her shoes off. For that she was paid a pittance. She was stoic – 'work was nothing to me, I couldn't care less'. Even so her life was hard. She regularly loaded and unloaded the boats herself and even recalled some of the loads – 25 tons of corned beef one run, 25 tons of timber the next. It was, in her own words, 'a hard life and a bad one' but she always said that she would not have changed it, and would do it all over again.

She was not alone in being hard worked from an early age. When he was between eight and nine years old Bob Bunn used to walk all the way from Braunston to Brentford, 93 miles, working all the locks and at the end of the trip was paid sixpence. Some children were working their own boats when quite young. One boatman was only fourteen years old when his parents, who were working a pair, acquired a third boat and put him in charge, with no pay. The children needed to be tough. Another boatman remembered cycling down the towpath to open a lock, when his bike hit a rut and catapulted him into the canal. Like many on the boats, he couldn't swim. But mother came past in the boat, hauled him out by his hair, without stopping and deposited him back on the bank.

Bringing up a family on a boat was never easy. When women went into labour, they had to rely on other boatwomen, instead of trained midwives, to deliver their babies. They then enjoyed the luxury of two days off, but after that it was back to work. Coping with the little ones was always a problem. When very small they had a harness that tied them to the boat, but they were soon running around. The only time they were kept strictly confined was in the tunnels, and no amount of pleading would get them out of the back cabin. 'Let's come out mam, let's come out mam – and mam kicked us back in.' Keeping clean was a special problem, when your home was also your workplace. Loading and unloading coal was the worst time, when the dust got everywhere and the children finished up as black as the cargo. Washing facilities generally consisted of nothing more sophisticated than a fire on the towpath on which the canal water could be boiled. Some places, such as Tring cutting, were always favoured, because the water was clear and clean. Hygiene was always something of a problem. Temple Thurston was very curious to know the answer to a question that he asked with delicacy: 'What do the women do when they are miles from habitation?' His boatman answered. 'Do?' said he. 'Why, look you sur – that hedge which runs along by every tow-path. If Nature couldn't grow enough leaves on that hedge to hide a sparrow's nest, it ain't no good to God, man, nor beast.'

Feeding the family had its problems, but there were always those who had their ways and means. One man was an expert on snaring rabbits, and thought it a poor night's work if he got

Washday on a boat; no washing machines and laundrettes in those days.

fewer than eleven from twelve traps. Others scooped moorhen and mallard out of the reeds with nets on their way past. And it has to be said that for many boating families, arable farmland was a handy back garden. Often fresh supplies, such as milk, were difficult to come by. Nick Hill recalled that a cup of tea was usually made with condensed or sterilised milk, which as a young boy, lock wheeling, he always called 'paralysed milk'. The greatest problem, however, was not providing food but preparing it, when you are also in charge of a moving boat. Many former boatmen have admitted that, in general, the women had the worst of it as far as work was concerned. They would do everything a man did – and 'women's work' as well. In most families, the man ruled, 'If a man said "Jump" even if they didn't think they could reach the side, they'd still jump'.

There were working pairs even before there were engines, and the first boat was the easier to steer, so that job generally went to the mother. It was not unknown for boats to be run right through the day and night, which meant meals had to be prepared on the move. Drinks were comparatively simple, just boiling a kettle in the back cabin to make tea or preparing a mug of cocoa. Then the mug would be dropped down at a bridge to be picked up by the following butty. Meals were more complicated, involving constant hopping to and fro between the range in the cabin and the tiller. Even when boats were moored for the night, the women had all the household chores on top of their work. They cooked the meals and cleaned the dishes, polished the brasses and blacked the stove so you could see your face in it. There were social evenings in the pub, but the men were first in and last out. It was simply the way things were and went unquestioned.

The men were also undeniably hard working and hard living. Almost everyone talks about the fighting – fighting about precedence at locks or more likely over an argument about nothing in particular in the pub. Everyone was agreed, however, that no one bore grudges. Two men who had fought violently one day would be the best of friends, sharing a table in the pub the next. And it was not just the men who got involved. One rather small and timorous boatman had a large and belligerent wife, more than capable of flattening any man who insulted her husband. Drinking and fighting seemed to be an inseparable part of the pub scene for many boatmen. One of these, who had better remain nameless, had a reputation for both activities. Many years ago he was involved with a film crew, working on a story based on the canals. They knew of his reputation and decided it was their mission in life to get this man with the prodigious thirst thoroughly drunk. He claimed that he could drink all the beer they could buy. Seventy-two bottles of Guinness later, the crew abandoned the attempt to see him under the table. Is the story true? I doubted it until I spent an evening in a pub in Braunston with the same boatman. I was very glad not to be paying. That night he told me a story about his most memorable fight. The cause of the argument had long been forgotten, but not the event. Things had turned very nasty, and his opponent finished up whacking him on the head with the metal lock key. He fell pole-axed. His no doubt inebriated companions were now seriously worried, convinced he was dead. What do you do with a corpse if you don't want to be had up for murder? You get rid of it. One of the party went off to fetch a length of canvas, needle and stout thread and proceeded to sew the body into a bag. Job complete they were about to launch it into the cut, when he woke up, mystified at finding himself in the dark and unable to move. The corpse suddenly became very active and noisy. Is that story true? Well, it is true as the tale of the seventy-two Guinnesses. But this is also a man who remembers that when he was a lad, his father stopped every Sunday morning and evening so that his wife could go to church.

Boating families were noted for their distinctive dress, which was another aspect of their lives that set them apart from those who lived 'on the land'. Danny Jinks called his best clothes his full rig. This consisted of a double-breasted waistcoat, shirts specially made by boatwomen, white cord trousers and hand-made shoes, never clogs. Belts were decorated by cord and were made by a saddle maker and he was quite envious of a cousin who had a pair of braces decorated by brass hearts and diamonds. In the nineteenth century, the women wore full-length skirts and blouses, which they often decorated with their own elaborate crochet work. The most distinctive feature was the huge bonnet, with a stiffened brim at the front and a frill, known as the curtain, hanging down at the back. At the death of Queen Victoria, the old white bonnets went out and were replaced by black, and the black bonnet lasted far longer than the memory of the queen. In later years, many boatmen wore trilby hats instead of caps, as better protection from the weather. There was always a danger of hats blowing off, but the secret for keeping it in place was 'wear a frown and screw it down'. It was not just the boatmen and women who got quite fancy clothing. The women used to crochet little bonnets for the horses to protect them from the sun.

It seems few people took much notice of the boating community until the middle of the nineteenth century. In 1846 a group of evangelical Christians formed the Incorporated Seaman's and Boatman's Friendly Society. They built missions on the Birmingham network, of which the sole survivor at Birchill is now a canal museum. It was originally 'The Boatman's Rest' and was well situated at the junction of the Walsall and the Wyrley and Essington canals, where in the early 1900s up to 200 boats a day went past. It offered an alternative to the pub, serving non-alcoholic drinks, providing games such as chess and newspapers for those who could read. For the many illiterate boat people it also offered a free letter writing service, which was much appreciated. Equally popular were the free washing facilities. The Brentford Mission had a lying-in room, which must have been a huge boon for pregnant boatwomen. A similar movement was started in Scotland. The Port Dundas Mission was founded in 1871 to provide for 'the social, spiritual and religious welfare of canal boatmen and their families'. The Salvation Army also ran boats, which among other things offered good quality second-hand clothes. Those who started and ran the various charities may have been concerned with the spiritual welfare of the families, but they were very well aware that it was the social benefits that provided the main attraction.

Charities did what they could, but officialdom took very little notice of the canal community for many years. Schooling was almost unknown. One exception could be found on the canal in Oxford. A local coal merchant, Henry Ward, bought a houseboat in 1839 and converted it into a floating chapel, which held services every Sunday, but also ran a school throughout the rest of the week. Conditions on the canals were eventually brought to the attention of the whole nation by the author of *Our Canal Population* published in 1878. Unlike many philanthropists, George Smith knew all about the miseries of child labour from first-hand experience. He was born at Tunstall in the Potteries in 1831, and by the age of nine he was working thirteen hours a day at a pot bank, carrying heavy loads all around the works. He found time to educate himself and rose to be manager of a brick and tile works. He began campaigning for the rights of the children in brickworks and thanks to his efforts an Act was passed in 1871, allowing inspectors to check on the conditions of women and children. He received the plaudits of many good citizens, but the owners of the works he managed took a very different view. He was sacked. It was then that he turned his attention to other groups of working children, the ones he saw regularly on the boats of the Trent & Mersey Canal.

Like many reformers he wanted to attract attention by dramatising his case and making the faults in the system as black as possible. In the process he was perfectly content to demonise a whole section of society:

> Utterly ignorant, as a large proportion of them undoubtedly are, of all religious knowledge, wholly without instruction, coarse and brutal in manner and entirely given up to the vilest debauchery and the grossest passions, can we expect, without extraneous assistance, that the children of such parents are ever likely to grow into anything better?

He estimated that out of a population of 100,000 on the canals, 95 per cent were illiterate, 90 per cent were drunkards and 40 per cent of the children had unmarried parents. The only low percentage was for members of a Christian church – 2 per cent. He described the children as being forced to live in accommodation 'scarcely the size of a gentleman's dog kennel'. It is a damning indictment, but has to be read with the ideas of reforming Christians of the period in mind. For them any person who went to a pub was, by definition, a drunkard. The fact that the canal pub was the one social centre where boat families could gather and were made welcome, until the foundation of the missions, was ignored. Church attendance was not easy to fit into the life of an itinerant, and the general attitude towards the boating community by the rest of society probably did little to encourage them to believe they would be welcomed. It was equally not uncommon to find communities that consisted of stable family units who had not gone through any formal marriage service – any more than it is today. The boat families might not have fitted comfortably with the way others thought they should act and behave, but that did not make them the degenerates that Smith described. If they lived in cramped quarters, that was from economic necessity, not from choice, and there is abundant evidence that the women kept their back cabins as immaculate as any bourgeois front parlour. Nevertheless there were evils in the system that he highlighted, not least the lack of schooling.

Staff and children at the Boatmen's Institute school, Brentford, in 1912: a girl on the left is wearing a typical boatwoman's bonnet of the period.

The old floating chapel for boating families moored on the Thames by the Oxford Canal in Oxford in the nineteenth century.

Smith began making his findings public even before his book was published and soon pressure was being put on the Government to act. The first result of the movement led by Smith was an Act that permitted local Sanitary Authorities to inspect boats for disease risks, which was admirable, though given the slum conditions that then existed in many industrial towns, the boats were probably less in need of those attentions than the homes of many other working people. School Attendance Committees were ordered to investigate the education of the children. 'Thank God! Thank God!' Smith wrote in his diary, 'Oh, how wonderful I feel'. Others were less enthusiastic. *The Times* expressed the true conservative view that the state should not meddle in personal lives:

> The floating home of the 'bargee' is to be invaded. Its privacy is attacked. Its liberty appears doomed to pass away ... with a display of inquisitorial power such as was never before dreamt of by any man conducting his boats through the canals and canalised rivers of England.

The Times need not have been so concerned. Although local authorities were given powers to act, very few bothered to use them. Smith continued to push for reform, greatly helped by the reception his book received. A second Act of 1884 was far more effective, for it established a centralised inspectorate, with real powers. The authorities were either very wise or extremely fortunate in their choice of the man for the job of making the system work. John Brydone cared passionately about his task, but realised that far more could be done by personal interest and persuasion than could ever be achieved by harsh enforcement. He built up a team who shared his philosophy and he was proud of the results:

> The boatman has realised that his position is not that of the despised member of society he once was considered ... The local inspectors recognise the importance of their influence upon the life and habit of the people, and it is a most remarkable fact that there is not one of the local inspectors against whom I ever now hear the slightest word of complaint.

However good the intentions of the inspectors, enforcement was always a problem. Theoretically the Act prevented overcrowding on the boats and ensured that all the children went to school. Checking the first part was never going to be easy in a shifting population. It was not unknown for a family with a lot of children to allow some to help out on other boats for a time for a suitable fee, but that did not mean that they would not be back again quite shortly. And if a family told an inspector that a child was just visiting from another boat, it was hard for him to disprove. Enforcing education was even more troublesome.

Boat children had an attendance book that was marked whenever they went to school. It was by no means uncommon for a child to appear, collect his mark and then be whisked away half an hour later because the boat was leaving. As far as the inspector was concerned, the book was properly marked and if there were enough marks it got his official stamp of approval. Even a child who stayed all day did not necessarily gain very much. The school in a village or town was organised around a curriculum designed for the local children who turned up every day, not for strangers who might drop in once or twice and never be seen again. Too many schools simply sat the boat children at the back of the class and left them with a picture book or drawing materials. At the end of the day they left knowing no more than they had when they entered. Many children did not even get this minimum attendance, and then the inspector wrote to the company. Most companies regarded one mark a week as quite sufficient, but even then there were families who preferred to pay a small fine, rather than lose the work of their children. Things improved in the 1920s when a special school was established on a converted

This group of children probably all shared the back cabin of the boat with their parents.

barge at Brentford. At least the teachers understood the special needs of the boat children and could adjust their lessons to their abilities. Literacy improved, but it was never possible to achieve the educational levels of children who received a more conventional education. In later years, families such as the Saxons deliberately gave up the boating life, precisely because they could see that their children would not have a future on the water, and they would only make a success on the land if they received a good education.

The one education all the children received was in the boating itself. Old boaters remember their first lessons in steering, standing on a stool so as to be able to see over the cabin roof. My own daughter did the same, simply because it was fun and exciting, and there's no reason that boat children were any less enthusiastic. There is, however, a world of difference between how our family coped with boating holidays on the canal, and the demands made of the working families. The most obvious differences are that we always had the advantage of a motor, had just the one boat to worry about and only ourselves to please, so we could travel as quickly or as slowly as we liked. Plus the far from unimportant fact that all we had to load and unload were suitcases and provisions – not tons of coal.

The motor is one of those things it is easy to take for granted, and just think of it as an alternative source of power, doing the job a horse once did. But in fact, it does a lot more. For example, coming up to a lock, you can stop the boat simply by reversing the engine. You can't put a horse into reverse, and get it to push a towrope instead of pulling it. It is necessary to unhitch the horse some way short of the lock, so that the boat can glide slowly through the open gates. It is then a matter of skill to get a rope round a bollard before the bows collide with the far gate. With a motor it makes little difference whether the canal is straight or bendy, but with a horse, the power supply is up front and slightly to one side, on the towpath. On convoluted canals or rivers, the steerer would find it very convenient to change the length of the towrope to suit the circumstances. By having the rope fastened to a stud on the cabin roof, it could be kept at hand and adjusted as necessary. All this required the skill and judgement that comes with experience.

The arrival of the motor solved some problems. Always when working a lock, the aim was to waste as little time as possible. Going downhill, for example, the boatman would leave the engine ticking over very slowly – the Bolinder was particularly handy for this. He would lift the paddles then jump back on board. Once the lock was nearly full, he would use the shaft to half drop the paddles, and then use the shaft to ease the gates open. With the boat inside, he'd drop the paddles and close the upper gates. A minute or so saved on a lock could add up to an hour saved at the end of the day.

Loading and unloading were jobs that required hard work, however you looked at things. But some cargoes were far worse than others. Grain was the worst. It was generally loaded loose, and anyone who has ever handled grain will know the clouds of dust that rise up as you move it. Once stowed, the problems weren't necessarily over. The grain was always liable to shift about during the journey, which could be dangerous on vessels with very low freeboard. After discharge, the hold had to be thoroughly cleaned out, because any grain left behind could start to germinate, and smell. Unloading was, in any case, far harder work than loading. It is a lot easier chucking shovel loads of coal down into a hold than it is throwing it up to the wharf from the bottom of a boat.

Those of us who were fortunate to see the last days of narrow boat carrying got a glimpse into a way of life that has ended. It was a life of hard work and difficult living, but it was also a life that offered a degree of independence and community spirit that was not to be found everywhere. It is also a way of life that has to be seen in context. At the beginning of this chapter, we looked at the changes in agriculture that might well have affected decisions to take up boating. By the middle of the nineteenth century, the Industrial Revolution had changed Britain forever but the contrasts between life on the land and life on the boats remained. In 1843, Dr G. Calvert Holland wrote an account of the life of the grinders of Sheffield, whose job it was to sharpen the edges of knives and the points of forks. They worked with metal dust flying all around them, and the vast majority died of throat and lung diseases. His statistics make terrifying reading. The worst hit were the fork grinders: 47.5 per cent died between the ages of twenty and thirty. Given the choice between working in a grinding hull or being out on a boat on the Sheffield & South Yorkshire Navigation there is no doubt over which was the better option.

10

A SECOND START

T
HE FIRST intimation of what the future had in store came to boatmen on the Aire & Calder in 1812. That was the year that the Middleton Colliery, just south of Leeds, embarked on a new venture. Coal was delivered from the pithead to the river in a conventional way on a tramway, using horses to pull the wagons. The Napoleonic Wars had resulted in big increases in the price of grain, so the colliery manager John Blenkinsop decided to try and replace the horses with something new. With the help of a local engineer Matthew Murray, he designed a railway locomotive based on a patent taken out by Richard Trevithick. It was a curious affair, using a rack and pinion system. This was not because they foolishly believed smooth wheels would not grip smooth rails, but to reduce the down force on the rails by giving extra traction. These were still the brittle cast-iron rails similar to those that had broken under the weight of Trevithick's first engines. It was the world's first commercial railway, and soon had many successors, though the cumbersome rack and pinion system was quickly discarded, and only reappeared many years later on mountain railways. Those who watched the first railways develop were aware that they represented a major change in the world of transport. If waterway transport was to keep its share of traffic improvements should be made wherever possible. Not surprisingly the Aire & Calder Navigation which had been there at the start was among the first to rise to the challenge.

The Navigation has a long history. In the seventeenth century the Aire was navigable by vessels up to 30 tons along the tidal river, from its junction with the Yorkshire Ouse as far as Knottingley, 16 miles from Leeds. In 1697 an engineer John Hadley was invited by the Mayor of Leeds and nine members of the Corporation to survey the river. He sent in his report and in 1699 the Act was passed, which made the Aire navigable to Leeds and also allowed the tributary Calder to be opened up to Wakefield. The locks were mostly built alongside existing mill weirs, and one can get a very good idea of just how the system worked by visiting Thwaite Mill in Leeds, a former putty mill which has been restored and is itself a fascinating place. The locks were of modest size, roughly 6oft long and 15ft wide, and vessels were soon covering the whole journey from the east coast to the heart of the two important wool towns.

What had seemed adequate in 1699 looked a good deal less so a century and more later. Wakefield and Leeds were thriving and there had been that little nudge in the ribs at Middleton to stir the authorities into action. First Rennie then Telford set to work on improvements, but by far the greatest change was the creation of what was in effect an entirely new canal from Knottingley to a junction with the Ouse roughly 3 miles downstream from where the Aire itself entered. It was here that the brand new port of Goole was created. The Aire & Calder Company were mightily proud of their new creation, and posted a notice in the *London*

The Middleton Colliery Railway, opened in 1812, was the first successful commercial railway in the world. It carried coal from the mine near Leeds to the Aire & Calder and marked the beginning of competition with the canals.

Canal improvements helped to fight off railway competition; this coal hoist at Goole lifted Tom Puddings up the tower, then upended them to shoot the coal into the hold of the waiting coaster.

Gazette in September 1828 extolling its virtues. The port of Goole was, they declared 'placed on a footing of equality with those of London, Dublin, and Liverpool, and of superiority of all others in the United Kingdom'. There might be a touch of Yorkshire exaggeration at work. In fact, there were two bonded warehouses on the wharfs, two docks and a timber pool, but the greatest pride was in the new 50hp steam tug the *Britannia*. What made the tug so very valuable was that it could meet ships off Hull, take them in tow as soon as they were cleared by customs, and bring them straight down to Goole.

Goole was a success, and the original ship and barge docks were later joined by a steamer dock and a railway dock. A new town grew up round this complex, but the nineteenth-century buildings have little of the elegance that one finds in the earlier canal towns such as Shardlow and Stourport. The original housing that survives has more in keeping with the terraces that were everywhere surrounding the new mills and factories of the north. The better-class houses have the rather drab solidity of Victorian villas. Public buildings, offices and hotels have a similar austerity. The big money went into the dock facilities. It is interesting to compare the cost of the improvements with those for the original works in the seventeenth century. The total cost then came to £25,000; the estimates for the nineteenth-century work were edging up towards £1 million.

Not everyone was immediately aware of what was happening. The first years following the success of the Middleton Colliery Railway seemed to offer little threat to the canal companies. The early lines were mostly based on the old tramways and linked collieries to navigable rivers, the Tyne and the Tees. Thomas Telford at the end of his life almost welcomed the arrival of the steam locomotive, but only as long as its role was limited to feeding cargoes onto rivers and canals. Even the opening of the first public railway, the Stockton & Darlington, in 1825 posed no threat. It was still, in essence, a colliery line, and although passengers were also carried, they did not enjoy the benefits of steam. The passenger coaches were no more than old stage coaches fitted with flanged wheels and pulled by horses. Others, however, saw the new line as something quite different. If a line could be developed to join towns, why not a line linking cities? The engineer responsible for the Stockton & Darlington, George Stephenson, was invited to look at a new route linking Liverpool and Manchester.

Once the proposal appeared in public, there was an outcry from the various canal and river companies. It was really like the Bridgewater story all over again. The newcomer would bring unfair competition to well established concerns and reduce them to penury. The only difference this time was that now it was the Bridgewater Canal Company joining forces with their old opponents, the Mersey and Irwell Navigation in opposition. The Parliamentary battle was every bit as fierce as any fought between river navigations and their allies the road hauliers against canals in the 1760s and '70s. It was ultimately just as futile. The railways would not be stopped, but they had to fight hard and were not given an easy ride. Even before they went to Parliament, the railway company had to lay out the proposed line, parts of which would inevitably go over private land. The railway needed to pass through the grounds of Captain Bradshaw, one of the Bridgewater proprietors; a perfect opportunity to create difficulties for the surveyors. Stephenson described his reception:

> I was threatened to be ducked in the pond if I proceeded, and of course we had a great deal of the survey to take by stealth at the time when the persons were at dinner; we could not get in by night, for we were watched day and night and guns were discharged over the grounds belonging to Captain Bradshaw to prevent us. I can state further, I was twice turned off the ground myself by his men; and they said if I did not go instantly they would carry me off to Worsley.

George Stephenson had worked his way up to his position as the world's first important railway engineer from a modest beginning, getting his first experience of working with steam when he was put in charge of a stationary engine at a pit head. He was a practical man, not a theorist. He spoke with a strong Geordie accent and was more used to dealing with engineering problems than legal niceties. But he was the man who had to go to the Parliamentary Committee to answer questions about the line – a line which had been surveyed in a great rush, in difficult circumstances and not very efficiently. The waterways army had brought up their heavy artillery in the form of their Counsel, Mr Alderson, and as Stephenson later noted, 'I was not long in the witness box before I began to wish for a hole to creep out at.' The cross-examination was merciless and every flaw was pointed out – and they were many. Stephenson had entrusted most of the work to assistants. He had to admit that he had not taken the all-important levels himself, and that many of them were inaccurate. He had proposed building a bridge over the Irwell which was not high enough to allow the vessels that used the river to pass underneath, and so it went on. Alderson, in his summing up, tore Stephenson, and his evidence, to shreds, 'I say he never had a plan – I believe he never had one – I do not believe he is capable of making one.' After that it was no surprise that the Bill was thrown out. It was a short-lived victory. A chastened Stephenson went back to work and the Liverpool & Manchester Railway received official approval.

This was arguably the great turning point in railway history. In their attempt to decide between two rival systems – one in which trains were hauled by cables between stationary engines and one in which they were pulled by locomotives – the Company ran a competition to find if a steam locomotive was up to the job. The competition was won by George Stephenson's son Robert, with his locomotive *Rocket*. It was to have all the important design features that were to be developed during the next century of steam railways. But there was also a surprise. The line was built with freight in mind, but there was a huge demand for passenger travel. Perhaps the canals, for whom passenger traffic was always subsidiary, might not suffer too much after all. Indeed, as the railway network spread, canals even picked up business carrying material to the construction sites. And even after completion many canals found business unaffected.

The mills and factories of the Industrial Revolution had developed alongside the canal system. As steam power gradually replaced water power, new buildings clustered close to the canals, which they relied on for their regular supplies of coal. You can see the change in many of the older industrial centres. In the Pennines on the Yorkshire–Lancashire border, the little hill-top village of Heptonstall was once a thriving centre of the woollen industry, based on weavers and spinners working at home. As mechanisation came in, the focus shifted to the streams and rivers of the valley to feed the water-powered mills. Walk along Hebden Water and you can see a typical building of the period, Gibson's Mill, but along the way you can also see the foundations of other, long-abandoned textile mills. The arrival of steam power shifted the centre once again. Now everyone wanted to be as close as possible to the newly opened Rochdale Canal and the town of Hebden Bridge thrived. With the valley floor all but filled with mills, houses had to clamber up the steep hillside, not in the familiar rows of back-to-backs but as unusual bottoms-to-tops. What appears as a conventional row of two-storey houses at the upper level is revealed as four storeys below, with one terrace sitting on top of the other. When the railway arrived, as it was bound to do, it closely shadowed the canal, but the cotton mills were still happy to take their coal supplies by water – as long as they received regular supplies it made no difference to them how long the coal took in transit. In the end, the only thing that mattered was cost.

There was a problem, however, that could only get worse. The railways were steadily improving. Engines were becoming more powerful and more efficient. The network was growing,

When Robert Stephenson planned his railway from Birmingham to London he took a line that closely followed the Grand Junction Canal.

Railways and canals did co-operate. This goods shed interchange linked the Great Northern Railway to the Regent's Canal at King's Cross, London.

with branch lines filling in the gaps between the trunk routes. The canals were static and unchanging, and unless something could be done their competitive position could only get worse. The Aire & Calder had shown the way: modernise. No one did more to lead the renewal movement than Edward Leader Williams, who was eventually to be knighted for his work.

His father, of the same name, was himself a distinguished engineer, who became chief engineer to the Severn Navigation. There were two sons. The younger, Robert, soon showed artistic talents and would eventually become well known as a painter and Royal Academician. Edward was born in 1828 and was educated at Worcester Grammar School. He left at the age of sixteen and was apprenticed to his father. In 1846 he briefly left the waterways to work as an assistant engineer under Joseph Cubitt on the Great Northern Railway, after which he was employed on harbour works around the country. It all helped him to get an overall view of Britain's transport system, from seagoing ships coming and going at the ports to the goods being moved around the country by water and rail. When he returned to inland waterways he had a very clear idea of what the challenges would be and what needed to be done.

In 1856 he was appointed to the post of chief engineer to the Weaver Trust, and at once began planning to ensure a future for trade on the river. The first essentials were to make it navigable by larger craft and to replace the old Weaver flats by modern steamers. He also saw that there was a valuable trade in salt that was being carried on the Trent & Mersey Canal that could usefully be fed into the Weaver. The only problem there was that the nearest approach came at Anderton, with the canal over 50ft above the river.

The locks on the Weaver had already been enlarged once – to 100ft long by 22ft wide and 10ft deep – but Leader Williams was not proposing just another small adjustment: he had something altogether grander in mind. The new locks would be 220ft by 42ft 6in by 15ft. That is a huge increase in capacity from 22,000 to 140,250cu.ft, and they were the biggest locks ever built in Britain at that time, so big that the enormous gates had to be operated mechanically using pelton wheels, a form of turbine. As a result 1,000-ton vessels could now use the waterway, but there remained an outstanding problem. The Weaver joined the Mersey at a point where the latter was notorious for shallows and difficult currents, so he decided to make a new junction further downstream. This involved making a cutting parallel to the Mersey shore for some 4 miles and creating a new dock complex at Weston Point.

Now he turned his attention to the salt trade. What was needed was a means of moving boats between the river and the canal. The idea of building locks was quickly dismissed and Williams went for a more dramatic option – a vertical boat lift. The idea was not entirely new. An Exeter engineer, James Green, was responsible for extending the Grand Western Canal, and to overcome the hilly country he did away with locks altogether and planned for one inclined plane and seven vertical lifts. They were comparatively modest and all have long since vanished, though extensive remains can be seen near Nynehead in Somerset. It is a bizarre sight as the canal at the lower level appears to come to a dead end at a high brick wall. That represents the most substantial part of the original lift to survive. Leader Williams was about to build on a far grander scale, and he called on the services of a consultant engineer, Edwin Clark, to take over the main design task.

The lift rose vertically above the Weaver and was joined to the canal by a short aqueduct. It consisted of two caissons each weighing 250 tons and able to take a single barge or two narrow boats side by side. These were counterbalanced and moved by hydraulic rams. The caissons were kept on track by means of attached cast-iron blocks running in grooves in the supporting structure. It was, and is, a remarkable structure, but suffered from the contamination of the local water by chemicals that rotted the sealants that protected the movable gates on the aqueduct and the caissons. That was temporarily solved by using condensed steam from the auxiliary engines in place of river water, but there was a major change in 1902. Electric motors were installed and instead of being counterbalanced by each other, each caisson was given its own set of counterbalancing weights. These hang off the caissons, giving the whole thing a curious Heath Robinson-like appearance. It has recently been fully restored.

The improvements were a huge success. In the 1880s, the Weaver was handling one and a quarter million tons of salt a year. Eventually the salt trade diminished – much of it being taken by pipeline. But there was a compensatory increase in traffic to the various chemical plants that lined the river. The plants may have done nothing but harm to the Anderton lift, but they ensured handsome profits for the Weaver navigation. The whole modernisation programme had more than repaid the money and effort that went into it. Soon Leader Williams was to be asked to consider a far more ambitious scheme.

Of all the transport connections in Britain, none has been the source of more important innovations that that between Manchester and Liverpool. First there was the Bridgewater Canal, then the world's first intercity railway, and now the citizens of Manchester were fretting again about the need for better and cheaper communications. There was no shortage of trade, but while all kinds of modernisation programmes were being pushed forward at Liverpool, Manchester felt left out. There was about as much good will between the two cities as there is between supporters of Liverpool and Manchester United today. Liverpool had been steadily extending its docks along the river, with the great engineer Jesse Hartley as the main driving force. He was responsible for the famous Albert Dock complex and many others, and by the latter part of the nineteenth century the docks had spread along the Mersey for nearly 2 miles. The river had been dredged and improved so that the largest ocean-going steamers could use the port. The trouble lay in what happened after that. Liverpool was not really interested in the upper stretches of the Mersey, and barges could be seen floundering about between mud banks on their way up to Runcorn. What was Manchester to do?

The drive for change was led by Daniel Adamson. There is an unlikely connection here with the story of the Liverpool & Manchester Railway. Adamson was born at Shildon, where the Stockton & Darlington Railway had its main engineering works. At the age of thirteen he was apprenticed to the man in charge, Timothy Hackworth. At the Rainhill Trials, designed to find a locomotive to run on the Liverpool & Manchester Railway, Hackworth had entered a locomotive of his own design, *Sans Pareil*. It failed to finish the course when a cylinder failed. As this had been supplied by Robert Stephenson's company at Newcastle there was inevitable talk of skulduggery and sabotage. The Hackworth family are convinced to this day that they should have won the contest. Adamson rose to be general manager of the engine works, but in 1851 he moved to a site near Manchester to set up his own iron works, specialising in manufacturing boilers. He would no doubt have received the wholehearted approval of his first boss at Shildon, as he now put forward a scheme that would put him in direct conflict with the railway company that had chosen the Stephenson design.

In 1882 he arranged a meeting at his home to discuss plans for a ship canal. The mayors of Manchester and Salford attended together with civic dignitaries, local industrialists, financiers and two engineers, Hamilton Fulton and Edward Leader Williams. They put forward two alternative plans for a ship canal to link Manchester to the Mersey: Fulton proposed a tidal waterway, which would be lock free. The idea was greeted with great enthusiasm and a new company Manchester Tidal Navigation was set up. It soon became clear, however, that though a tidal canal would be comparatively cheap to build, a more conventional canal with locks as suggested by Leader Williams would be a far better option. So the Manchester Tidal Navigation became the Manchester Ship Canal – or would do so as soon as Parliament approved. The scene was set for the third battle of vested interests, and this one was to prove more protracted

The greatest canal engineering feat of the late nineteenth century was the Manchester Ship Canal, which used specially laid railways during construction.

The train of barges looks minute compared with the other vessels on the Manchester Ship Canal.

– and a good deal more expensive – than its predecessors. On one side were the Manchester men and against them stood the Liverpudlians and their allies. The last fight had been between the Bridgewater Canal and the railway, but now they were united against the common foe. Two Bills were thrown out in the long and protracted arguments, from which at least the lawyers prospered. It was estimated that legal and other fees amounted to £350,000 – which was more than ten times the amount spent by the Liverpool & Manchester in getting their railway bill passed. The owners of the Bridgewater Navigation Company did pretty well too. As part of the agreement they were bought up by the Ship Canal for £1,710,000. Once this was added to the construction costs, the total amount needed came out at a whopping £8 million. Finally, just over five years after that first meeting at Adamson's house, the first sod was ceremonially cut and work could get under way.

This was to be one of the greatest engineering projects that Britain had ever seen. The days when canals were all dug by hand were long gone, but even so when the work was at its most fevered there were 16,000 men and boys employed and 200 horses. The figures might suggest that little had changed since navvies first began digging canals, but other statistics tell a different story. There were also a hundred steam excavators of various kinds and a couple of hundred cranes. The vast amount of material that had to be brought in and the spoil that had to be shifted out was carried in railway trucks, running on 223 miles of temporary track. Altogether 173 locomotives were in use. Statistics can be boring, but looking at the ones for this canal the numbers are simply mind boggling. The official handbook for 1894 gives pages of them and here are just a couple to give at least a notion of the scale of the enterprise: 76 million tons of spoil was removed, a fifth of which was rock and 175,000 cubic yards of bricks were used. No one actually counted them, but if they were standard English bricks that works out at nearly 76 million bricks.

Fortunately we do not have to rely just on statistics to get a picture of the workings, for unlike the earlier canals, this was the age of photography. There are dramatic shots of the excavations showing trains of wagons stretching far into the distance, each apparently being loaded by three or four navvies. By contrast the giant excavators were usually shown in an almost empty scene, but almost dwarfed by the immensity of the excavations. This was a canal designed for vessels up to 80ft beam, and in places it was up to 30ft deep and over 200ft from bank to bank. That is a very big trench to dig. Special funicular railways were constructed in some parts of the excavations to move carts up and down the steep banks, and the banks themselves were stabilised by driving in wooden stakes which were then interlaced with willow branches.

The pictorial record does more than merely recall the technical details of the work, for there are glimpses of the social life as well. The men and their families lived in temporary accommodation of neat little houses made of corrugated iron – 'tin towns' they were called. Everything was quite crude. Open ditches ran either side of the muddy main street, with planks laid across them to reach the front doors. They may not have been ideal, but they were a vast improvement over the shanties of earlier workings, and were real family homes. There are also glimpses of the two worlds. One photograph shows a train with the directors' viewing coach. The little tank engine is totally basic, without any cab to shelter the crew, but the coach is all polished wood and a director in shiny top hat and frock coat is shown waiting to board. He could easily be taken as the original model for the Fat Controller, though the tank engine bears no resemblance to Thomas. Even grander transport was available once the canal was open for business. The Chairman had his very own yacht, not some overgrown dinghy, but a two-masted steam yacht of considerable grandeur.

Apart from the sheer scale of the works, there were a number of specific problems to solve. Firstly, all the bridges across the canal had to be movable – tall-masted sailing ships were still very much an important part of the maritime world. Hydraulically operated swing bridges were not new: the most famous early survivor crosses the Tyne in Newcastle and was built by the Armstrong works in 1876. But there was another obstacle in the way that presented a brand new problem. The canal had simply absorbed the River Irwell, including the section crossed by the Bridgewater Canal at Barton. Brindley's aqueduct was far too low and had to go, but what could replace it? Leader Williams' solution was, literally and metaphorically, revolutionary – a swing aqueduct. The iron trough, 18ft wide and 235ft long, is mounted on roller bearings on a central pier. The entire trough can be closed off by gates with rubber seals and swung while full of water. With traffic virtually gone from the ship canal it is rarely moved these days. Thirty years ago I was happy to be stopped on a journey down the Bridgewater and forced to wait while the aqueduct was moved. What struck me was the smoothness of the operation, which scarcely seemed to cause more than a ripple on the surface of the water as it moved. If Brindley's pioneering aqueduct had to go, then it could not have found a more spectacular successor.

Manchester finally had a direct access to the sea, and a whole new dock complex was built, which in time would have 200 acres of basins and over 5 miles of quays. It at once attracted local developers to set up what was to be the forerunner of the modern business park, Trafford Park.

The message that broad canals had a viable future, whatever the fate of the narrow canals might be, spread across the country. The Gloucester & Sharpness Canal was a long time in the building. The Act for the 17-mile-long canal was passed in 1793 but it was not finally opened until 1827. A mere fifty years later, the dock at Sharpness was already proving inadequate to cope with the far larger ships coming up the Bristol Channel, which were often left at anchor in the tidal river, waiting for a space to be freed in the dock. In 1871 work began on a whole new dock complex and a new entrance lock. The old lock took vessels 81ft by 19ft 6in, which must have seemed massive in 1793; the new allowed 163ft by 38ft. New warehouses were added at both Sharpness and Gloucester, and though the trade on the canal declined, Sharpness retained its importance, and is still in commercial use. There is also a thriving ship repair business, based on the dry dock.

Other broad canal and river navigations were improved in the latter part of the nineteenth century, including the River Lea. It was here that Britain's first pound lock was built, right back in the sixteenth century, and today it has a new role, providing a transport route for construction material for the London Olympic Games site.

The narrow canals presented rather more of a problem and it was no easy matter to make real improvements, though there were some areas that could be tackled. The inconvenience of Brindley's wandering, wavering contour canals was becoming more and more troublesome by the 1820s, and nowhere was this more aggravating than on the all-important routes through Birmingham and down into London. Telford was called in to look at the problem. The first idea to be acted on was to create a new Birmingham main line that would slice, straight and true, through the Brindley wriggles. The old canal would not need to be closed, and indeed there were good reasons for keeping it open, for there were important customers along its banks. Instead the curves would be retained as a series of loops. The new line was to represent the very best in canal technology. Nowhere can this be seen more clearly than in the deep cutting at Smethwick. To span it a magnificent new road bridge was designed by Telford. Galton Bridge is a single cast-iron span, which at the time, was one of the longest yet built. It is unostentatious and makes its effect by its functional simplicity and gentle curves. Telford also had another problem. His proposed route cut across the line of a branch from a canal made by John Smeaton in an earlier attempt to improve on Brindley. As the two canals were at different levels, he needed an aqueduct to cross his cutting. This time he decided to let his architect self-rule over the engineer. He had developed a taste for the fashionable Gothic, and this aqueduct was to get the full treatment. The Engine Arm aqueduct is a genuine oddity, the basic iron structure being decorated with rows of pointed arches that look as if they might have been borrowed from a Victorian church. But time has lent it a certain charm.

Telford also turned his attention to the rest of the route from Birmingham to London, and here he came up with a far more radical proposal. He wanted to build a brand new wide canal with broad locks and a towpath to either side. It was planned to join the Grand Junction near Braunston, which caused panic in the ranks of the Oxford Canal Company. All the valuable traffic that used their canal north of Napton would simply vanish onto the new route. There was only one possible solution. They had to modernise – and quickly. Within a week of seeing the announcement of the proposed London & Birmingham Junction Canal they had appointed an engineer to prepare a plan. The man chosen was Marc Brunel, an engineer best known, if very unfairly, as the father of Isambard Kingdom Brunel. A recent biography of Marc was given the title – *The Greater Genius?* The question was meant to be taken seriously. Brunel acted quickly and produced a rough plan, took his fee and left. The task of providing a detailed plan went to an engineer who, although he was only thirty-five years old, had already had a remarkable career. Charles Vignoles was born in Ireland into a military family and, although for a time it looked as if he might be heading for a career in the law, he finished up in the army during the Napoleonic Wars. Like many young officers he was pensioned off when the war ended and put on half pay, which was not enough to live on. He promptly set off for America, with the original intention of joining Simon Bolivar's army, but never got further than South Carolina. Instead of fighting, he finished up working for the state civil engineer, then went to Florida as a surveyor and mapped the state. The work was not very well paid, and in 1823 he returned to England, working as a surveyor, and soon had his own London office. The Oxford Canal was his only work on the waterways, for like most engineers of his generation he was soon to be swept up by the new enthusiasm for railways.

The Brunel and Vignoles plan resulted in the distance from the junction with the Coventry Canal to Braunston being shortened from 41 miles to 27½ miles, which gives some idea of just how meandering the original line was. Once again the new canal cut straight through the old, leaving a series of loops. Unlike the Birmingham, the northern Oxford passed through no great industrial areas, and the various loops were gradually abandoned. It is still possible to take walks in the area and come across a canal bridge in the middle of a field, with only a slight indentation to indicate that there was once a waterway there as well. One of the biggest advantages of the new route was the replacement of the old, narrow tunnel at Newbold on Avon with a new version, higher, wider and complete with towpath. Some years ago, I was photographing the old tunnel, down by the church, when a local wandered up and asked if I knew what it was. Curious to know what he might say, I asked for the explanation. He solemnly informed me that it was an air-raid shelter specially built into the hill in 1939. Which all goes to show that local knowledge can be less than entirely reliable. The measures taken by the Oxford Company did the job; nothing more was heard of the London & Birmingham Junction Canal. There were to be further changes to the Oxford. In 1840 the locks at Hillmorton were doubled, a set of three new locks being built alongside the original. It all helped to keep traffic moving.

Nothing else of very great note was done to improve the narrow canals until the very end of the nineteenth century. The most ambitious scheme was devised to remove a notorious bottleneck on the Leicester section of the Grand Junction at Foxton. Here the steep hillside was attacked by two five-lock staircases set very close together, but with enough room for boats to pass between the two sets. Even so they caused a great deal of frustration. What was even worse

The Foxton inclined plane was an attempt to bypass the nearby lock staircase. Here the engine house is seen under construction.

was that these were narrow locks, where the rest of the route to Market Harborough had broad locks. Fellows, Morton & Clayton pushed the company to make improvements, so that they could use the canal for barge traffic, and as they were the canal's most important customers the company obliged. In 1898, work began on the Foxton inclined plane. It consisted of two counter-balanced caissons, each able to take either a barge or a pair of narrow boats. The caissons ran sideways along a railed track on a 1:4 slope, moved by wire ropes and a steam winch. Under the new system a pair of narrow boats could be going up the slope, while another pair came down, and the whole trip took only twelve minutes. Under the very best conditions, a single narrow boat took forty-five minutes to pass through the locks. And, of course, there was only the one passing place, so a boat could easily be stuck in the middle, waiting for a turn.

The advantages seemed to be huge, but in practice there were technical problems, and most boatmen were unwilling to pay the extra fee for using the incline. Opened in 1900 it was closed ten years later. No one could claim it was a great success. Other more conventional improvements were on their way, however, for nearby canals.

Not all the problems involved in improving traffic flows on this important London-Birmingham network had been solved by the Oxford improvements. The Grand Junction had been built with broad locks as far as Braunston, and although the Oxford had been improved, the more direct route to Birmingham was still via the narrow locks of the Warwick & Napton and Warwick & Birmingham canals. Nothing was done about this problem until 1919, when the Grand Union Canal Company was formed from eleven earlier companies, including the Grand Junction, the Regent's and the connections up to Birmingham. They set about a major modernisation programme, with a million pound budget, which included building new locks. The most impressive example of their work can be seen at Hatton on the former Warwick &

Birmingham. A flight of twenty-one broad locks, known rather fancifully as 'The Stairway to Heaven' lifts the canal 146ft in just over 2 miles. The old narrow locks were not wasted – they formed overspill weirs. The first thing you notice when you approach them – apart from the awareness that there's a lot of hard work lined up in front of you – is the paddle gear. This is all enclosed to protect it from the weather and helps to make this one of the most distinctive flight of locks in Britain.

In the end, there was only so much that could be done to improve the narrow canals, short of a complete rebuild. The capital invested in the broad canals paid off: the narrow looked doomed for eventual extinction. That might well have been the case if they had not found a new use. Right at the beginning of the canal age the pamphlet promoting the Trent & Mersey had rhapsodised about the delights of canal travel. At the very end of many pages explaining the advantages of moving freight by water, a paragraph or two touched on another aspect of the canals:

> And if we add the amusements of a pleasure-boat that may enable us to change the prospect, imagination can scarcely conceive the charming variety of such a landscape. Verdant lawns, waving fields of grain, sequestered woods, winding streams, regular canals to different towns, orchards whose trees are bending beneath their fruit, large towns and pleasant villages, will all together present to the eye a grateful intermixture of objects, and feast the fancy with ideas equal to the most romantic illusions.

It may have been an afterthought, but it was to prove a prophetic statement.

11

PASSENGERS AND PLEASURE

P ASSENGER BOATS had been used on rivers long before the canal age began. In London, the Thames was both a barrier and a main thoroughfare. There was a steady procession of boats going to and fro between the south and north banks as well as up and down the river. The old London Bridge looked very picturesque, with its rows of arches and its topping of shops and houses, but it was hopelessly overcrowded. If you wanted to cross the river, it was often far quicker to hop on a boat, and fares were low. A price list of 1708 gave a charge of just twopence to get from London Bridge to Limehouse. The small boats, were simply known as 'oars' and were the equivalent of the modern London taxi. The watermen used to shout out 'oars, oars' to attract customers, and one Frenchman noted in his diary how he had misunderstood the cry and was disappointed to be offered a boat ride rather than a lady of easy virtue. On great occasions, the river was crowded. In 1662, Samuel Pepys tried to hire a boat in order to see the arrival of Charles II and his queen from Hampton Court. But even though he offered to pay 8s there was nothing to be had, and he watched the scene from the top of the Banqueting Hall. 'Anon came the King and Queen in a barge under a canopy with 1,000 barges and boats I know, for we could see no water for them.'

Travel on the tidal Thames was not always a pleasant experience, and where the Fleet emptied into the river it was downright disgusting, as Ben Jonson vividly described.

> Which, when their oares did once stirre,
> Belch'd forth an ayre, as hot as at the muster
> Of all your night-tubs, when the carts do cluster,
> Who shall discharge first his merd-urinous load.

The upper Thames was a good deal more pleasant, but considerably more dangerous. In the age before pound locks became common, the boats had to ride the flooding waters over the flash locks. In 1634, a passenger boat overturned in the Goring flash, with the loss of over sixty lives. By 1760, flash locks were happily rare and river boats were well able to compete with road coaches for trade. The canal age had scarcely begun before canals began offering a similar service.

Once again, the Duke of Bridgewater led the way in 1772 with a service between Manchester and a pier on the outskirts of Warrington. He started with two boats, one carrying 120 passengers and the other eighty, and soon the service was extended to both Runcorn and Worsley.

A whimsical view of
the packet boat on
the Grand Junction.

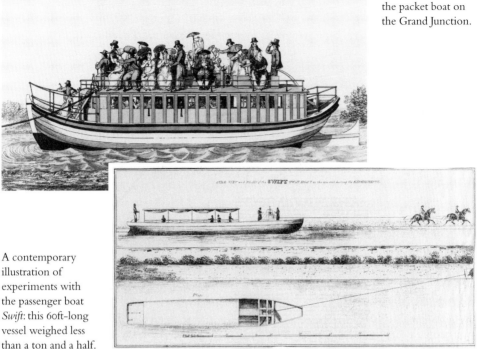

A contemporary
illustration of
experiments with
the passenger boat
Swift: this 6oft-long
vessel weighed less
than a ton and a half.

One reminder of those days is Packet House, Worsley, a typical black and white building of that
area, with steps in front from which passengers boarded. The original boats had three classes,
with prices ranging from 2s 6d for first class to 1s for third. Sir George Head took a trip through
the manufacturing districts of Lancashire in 1835, starting his journey on the packet boat from
Runcorn. There had been a few changes in the years since the service started. It had just the
two classes, but was described as tidy and clean. The passengers could either sit in the cabin, or
enjoy the fresh air from benches on the roof. He described the experience of being a passenger
as quite pleasant, 'There he sits, without troubling himself with the world's concerns, basking
in the sunshine, and gliding through a continuous panorama of cows, cottages and green fields.'
He became less enthusiastic as they got nearer to Manchester, where the canal became 'black
as the Styx, and absolutely pestiferous'. Now that cows had given way to factories, he began to
take a rather more jaundiced view of this mode of travel.

He described the boat as being towed by a pair of 'clumsy cart-horses', ridden by a pair of
boys, no more than twelve years old, who whipped the beasts along, with a good deal of energy
and very little skill, 'Half the strength of one horse was continually exerted to prevent itself from
being dragged into the canal by the other.' Even so, he estimated that they made about 5mph
– not bad for a pair of old carthorses. When he tired of watching the scenery he could enjoy
a substantial meal, and for a shilling he feasted on sirloin of beef with fried onion, followed by
salad and cheese. There were other supplies on board, which passengers took advantage of:

A woman contrived to pitch herself head-foremost off the top of the platform where she was
sitting, down upon the deck. She fell with such violence that I really thought she might have
been killed. As it was, she was not hurt, and as I picked her up, she set forth a sigh, which smelt
so strongly of rum that I was happy to consign her collapsed form into other hands.

Packet boats were soon to be found on canals across the country, running to regular timetables. These vessels were given priority, and any other craft meeting them had to drop their tow-ropes to let them through. Failure to do so resulted in a hefty fine. The Duke of Bridgewater's boat offered a rather more rapid response – they carried sickles in the bows to slice through the offender's line. The packet boats carried a few parcels as their only cargo. By contrast, the market boats carried produce and a few passengers, on a less strict schedule. They were far cheaper than the packets, but offered only rough accommodation – often no more than a perch on top of sacks and crates. The poor would sometimes come to a private arrangement with boat crews to get a ride for a few pence at most – an early form of hitch-hiking. As with modern practice it was not without its risks. In 1839 a young woman called Christine Collins managed to get a lift on a Pickford's fly boat in Preston, so that she could join her husband in Liverpool. She never arrived. On the way she was raped, murdered and dumped in the canal. There was the inevitable public outrage and a general condemnation of the boating popula-tion, even though it was an isolated incident and literally millions of people were passengers on the canals without coming to any harm.

The arrival of the railways posed a far greater threat to passenger travel on the canals than it did to the freight trade. By the time Sir George Head made his trip on the Bridgewater Canal, there were already over 5 million passengers a year travelling by rail. Why spend a day travel-ling between Liverpool and Manchester by canal when you could do the same trip by rail in about an hour? One of the first canals to try and improve services was the Glasgow, Paisley & Johnstone, a rather late arrival on the scene that opened in 1811. They were operating three slow package boats, and like all canal owners were reluctant to speed things up because of the damage done by the wake of a fast boat. Things might have stayed that way, but for an accident on the canal, described by the distinguished Scottish engineer and ship builder, John Scott Russell:

> A spirited horse in the boat of William Houston, Esq. one of the proprietors of the works, took fright and ran off, dragging the boat with it, and it was observed to Mr. Houston's aston-ishment, that the following stern surge which used to devastate the banks had ceased, and the vessel was carried on through the water comparatively smooth.

Russell went on to investigate the theory behind the phenomenon, but Houston was just happy to accept the remarkable fact that a canal boat could travel at speed without harm to the fabric of the waterway. He went on to design what came to be known as Scotch or swift boats. The first were almost ludicrously slender, with iron hulls 60ft long but only 4½ft beam. Later they were extended to 70ft by 6 to 7ft. They were pulled by two or three horses, managed by liveried postilions, and could achieve speeds of up to 12mph, though they had to have frequent stops to change horses. They soon became popular on other waterways and Sir George Head, continuing his northern canal journey, took a Glasgow built swift boat, and eyed it with a certain apprehension. Unloaded, and riding high in the water, the slender craft seemed danger-ously unstable, 'She was to all appearance, so cranky – toppling and rolling from side to side so awfully when empty, that people took a panic, and many declined on any account to venture.' Once a few brave souls had clambered on board, she settled down in the water and the rest were persuaded to board. It was an exciting trip, with two horses, the postilion riding the rear and controlling the leading horse by reins and a long whip. He, himself, was a resplendent char-acter in his colourful company livery. The horses were soon off at a full gallop, with the slender vessel moving smoothly through the water behind them.

Passengers on the swifts paid a premium for the privilege, but they were treated accordingly. A traveller described the cabin in the stern as having a table and comfortable chairs, books to

An illustration from *Two Girls on a Barge*, an early account of a pleasure trip in a converted narrow boat. Here is the party at Foxton.

read and a fire in winter. They were also assured of good food and wine on the journey. It must have been a very pleasant form of travel and became popular with excursionists.

The swifts were little more than a rearguard action in the face of the inexorable spread of the railway network. By 1845 the number of railway passengers had leaped to 30 million, but they also provided an opportunity as well as a challenge. The opening of the Shrewsbury & Chester Railway provided just such a useful opening for the Montgomery Canal. They did a deal whereby the canal company ran boats to a fixed timetable from Newport to connect with trains at Rednal. It was a demanding schedule, which allowed little more than five hours for the journey of 32 miles and twenty-two locks. It was a success and in the first year nearly 1,000 first-class passengers were carried and nearly 7,000 second class. Sadly it did not last, but went the way of other swift services. It was a contest the railways were always going to win. The closure of this service may have marked the end of traditional passenger services, running to a timetable. The railways were undeniably quicker, but not everyone was in a hurry. There was a hint of the way ahead in the north-west of England. The Romantic movement had made the Lakeland hills into a number one tourist attraction, and the Lancaster Canal's swift boats offered a splendid way of getting from industrial Lancashire right up to Kendal at the foot of the hills. Boats attracted tourists who could enjoy the experience for its own sake and who preferred the clatter of hooves to the rattle of the steam train. By the latter part of the nineteenth century, excursion steamers became a popular attraction on the broad canals of Scotland. Elsewhere a new vogue was very slowly getting under way – do it yourself tourism, at first limited to Church and Sunday School outings. The rector of Camerton near Radstock arranged just such an outing in 1823, which seems to be the first recorded example of such a trip:

> Having engaged one of the coal barges, I had it fitted up for the ladies with an awning and matting against the sides, and tables and chairs from the public-house, in which we proceeded about eleven o'clock to Combe Hay, where we visited the Mansion House, walked round the premises, and afterwards dined under the trees near the cascade.

It is not an experience that can be repeated today. All that remains of the canal in water is a short stretch where it joins the Kennet & Avon by Dundas aqueduct. But you can walk the line and enjoy the magnificent scenery, and hunt out remains of the old canals, from ruined locks to traces of the inclined plane near to the spot where the vicar and his party had their picnic. What sets this trip apart as being something special was the boat, not a specially built passenger vessel, but just an ordinary canal boat adapted for pleasure boating.

There was also a growing interest in travelling waterways for pleasure in privately owned boats, particularly rowing boats and punts. *Three Men in a Boat* is the most famous example, describing a trip up the Thames, and most accounts favoured rivers, though by the end of the nineteenth century a series of Oarsman's Guides appeared that also included canal descriptions. A book of 1873 was based on a canoe trip from Leicestershire to Greenhithe, part of it by canal. A far more ambitious scheme was described in a very enjoyable book published in 1891 – *Two Girls on a Barge* by V. Cecil Coates. The 'barge' was in fact a narrow boat and came complete with 'Mr. and Mrs. Bargee'. Quite what the traditional boaters made of their temporary guests is sadly not recorded, but it probably made for pub conversations for years to come. The first task was to make the boat fit for habitation by the young ladies. A carpenter was employed to divide the hold up into cabins and a saloon. The girls were not into Spartan simplicity, 'There were curtains to be hung, Liberty curtains that had taken a whole day to choose, and "dhurries" to be draped over the fresh-scented pine of the little cabins; and Liberty again in innumerable hangings to be arranged all round the bulwarks gracefully.' Finally they were ready to set off in their floating Victorian parlour. The girls themselves dressed for the occasion looking as if they were ready for a refined garden party rather than a working trip, and they were joined for part of the journey by the brother of one of them, decked out in striped blazer and straw hat. There was no indication that anyone considered helping out Mr and Mrs Bargee.

The narrative is generally rather fey, and there is not really a great deal about the canals themselves, up the Grand Union and the Leicester Arm, apart from a long description of going through Blisworth tunnel. The author gives a somewhat melodramatic account of travelling through the 'black, domed passage' where the damp stones 'played fantastic tricks with the imagination' and 'every sense became distorted, unnaturally acute.' The silence of the tunnel was said to be 'appalling'. The boatman clearly had a sense of humour and decided that this was just the time to make his mildly hysterical passengers even more nervous. He told them of the great White Spectre that haunted the tunnel, and how anyone who saw it was doomed. 'Ye just slips your foot, or overbalances, and the black waters swallow ye.' At this point they met another boat, where one of the leggers was a young boy, clinging tightly to the board. The boatman found the sight just the thing to embellish his tale. The boy was 'well-nigh skeered' for his father had seen the Spectre just the day before 'so then he slipped, and the black waters swallowed he'.

The experience of the two girls was very different from that of the best known of all the early canal travellers, E. Temple Thurston, on board the *Flower of Gloster*. His book begins with an account of being extremely irritated by the over-informative official guide in St Paul's Cathedral, after which he decided that he wanted to visit places where there were 'no guides and scarce a map' and he could make his own discoveries. The canals seemed to provide exactly what he wanted. So he hired his narrow boat, which he certainly did not deck out in Liberty prints, and a boatman and set off on a journey from Oxford that would eventually deposit him back on the Thames at Lechlade. His first impressions were not very satisfactory as he passed along the canal between Oxford and Wolvercote through suburbs, which he describes as being the work of 'jerry-builders – men of execrable taste'. How astonished he would be to know that these houses he condemned, their gardens sloping down to the water, are considered to

P. Bonthron and friends on the Oxford Canal: his book is the earliest account of travelling by motorboat on the canals.

be among the most popular properties in Oxford. He only really cheered up once he had reached Thrupp. It was here that he finally fell in love with canal travel and the slow progress through a rural landscape. His is a very English point of view: a dislike of the modern and a preference for the old. It is still a view you hear today, except, of course, that the old we admire is often the modern he detested.

Where Thurston differs most markedly from the girls is in his attitude to the boatman he had hired, with whom he shared the back cabin. He had a name, Eynsham Harry, and the author makes him appear, not as a crude illiterate boatman, but as a wise countryman. 'When a man makes you think,' he wrote, 'there is more than just something to him.' So they talked about everything, but mostly about nature, where Harry proved to have an intimate and profound knowledge of the wildlife that they met along the way. Thurston was impressed. '"It comes to this," said I, "that you're the true-born philosopher. You learn your philosophy from the hedgerows. I only play at it."' It could seem patronising, but isn't. The other marked difference is the interest he showed in the canal itself and its working life – as long as it remained limited to a rural environment. There were he declared two places that he regarded as hells on earth – Monte Carlo and the Black Country. So for him, the canals remained interesting only for as long as they were removed from all traces of the industrial world which they served. It is an attitude that is still common today, when canal holidays are presented as 'get away from it all' experiences.

In 1916, a mere five years after Thurston's book appeared, another account of canal travel appeared – P. Bonthron's *My Holidays on Inland Waterways*. There could scarcely be a greater contrast. Where the former had been content with a traditional horse-drawn boat, taking his share of the work and revelling in the slow pace, Bonthron travelled by motor boat and prided himself on covering as many miles as humanly possible on every single day. A typical entry reads: 'We were really rather pleased with our run on that day, as all together we had done 25 miles over part of the three canal companies' properties, and the very large number of 66 locks – an excellent piece of work.' Again, the attitude is far from unknown today. The start of the book sets the tone for what is to come:

'All clear ahead?' 'Ay, ay, sir,' responded the chief mate, giving the starting handle a turn and letting go the painter, in compliance with further commands. So now we're off at last on our long-talked-of canal cruise through the heart of England in our 6 h.p. Daimler.

They were the traditional three men in a boat, but with the addition of the motor boat engineer and the handy man. Eventually he travelled a very large part of the network, from the Caledonian in the north to the Royal Military Canal in the south. The accounts are perfunctory, but there are snippets of information that prove fascinating. The author grumbled about all the letter writing that was needed to travel the system, where permission was needed from each individual canal company, though he found all of them helpful. It was an interesting experience on both sides. Motor boats had yet to make any real impact on the canal scene – Bonthron described himself as a pioneer – and the sight of a 27ft launch must have come as a surprise to many. Already some of the canals were so weedy that they found it impossible to travel by motor. One of these was the Kennet & Avon, and they took a skiff, on which they were able to set a simple lugsail when there was a favourable breeze. Their first day's travel was from Bath to Devizes, which certainly bears out the author's description of the crew being enthusiastic oars-men. They did, however, cheat at Devizes, where the skiff was pulled out of the water and taken past the twenty-nine locks on a flat cart hauled by a donkey. Along the way, he mentioned that he thought it would be a very popular business to run pleasure cruises on the lock-free section of canal above Bath locks. He was quite right.

Altogether Bonthron travelled some 2,000 miles of waterways, which included many rivers as well as canals, and his book was largely intended as a guide to encourage others to follow after him. It gives details of distances and recommends hotels for overnight stops. One other canal traveller, of the same period, preferred to take his own accommodation along with him. C.J. Anderton's 1916 book is called *A Caravan Afloat* and that is a good description of the vessel. It was a high-sided boat 36ft long and 6ft 6in beam, which could be bow-hauled, poled or moved by a paddle wheel. It did not, however, have a motor – the wheel was worked via a cog and chain, like a bicycle. He described the experience as being like cycling up a very steep hill and the driving position as 'a seat of torture'. He was hoping to persuade others to buy these cumbersome little craft at the modest price of £40 for a bare hull; there do not seem to have been many takers.

Far and away the most interesting, and certainly the most important, account of travel on Britain's canals is L.T.C. Rolt's *Narrow Boat*, published in 1944. Tom Rolt took his converted narrow boat *Cressy* on a trip round England's canals. Unlike all the earlier writers, he was fas-cinated by the working life of the canals and the industries they served. Where others had seen only filth and noise, he saw high drama. It is not all that long ago that it was still possible to take a boat on the Trent & Mersey that pierced the heart of the Shelton steel works in Stoke-on-Trent. Anyone who did so will recognise Rolt's description of the experience and I, for one, felt just the same sense of drama and excitement. It is worth quoting at length:

'Cressy's' white windows, that for so long had seen unfold before them a slowly moving pattern of field, hedgerow and tree, now looked directly into a clangorous rolling mill, lofty as the nave of a cathedral, where white-hot billets of steel were being flattened as easily as pastry under a rolling-pin, or grappled by the electric cranes which rumbled high overhead. Workmen, their face streaked with sweat and grime, looked up from their task of feeding the rolls to grin and nod, while at one point where the crane track projected over the canal a crane driver leaned from his cabin directly above our deck to call after us, 'What about a trip?' A damp white mist shot through by the sunlight with miniature rainbows momentarily enveloped us as we passed the cooling towers, and beyond these the coke-ovens were belching steam and flame alternately. Opposite them, towering above us, reared the fiery heart of this monstrous organisation – the blast furnaces. Lifts were creeping up and down their pitiless steel sides feeding them with fuel, and, as we passed, one of the cones that close their throats was lowered to admit a charge, the air above shimmering in the sudden blast of intense heat which shot skywards.

The frontispiece from L.T.C. Rolt's iconic book *Narrow Boat*.

The title page from one of the series of waterways guides published by British Waterways Board in the 1950. Each book was illustrated by woodcuttings.

The works have long gone, but for some, including myself, the reminders of the fast-disappearing industrial world are an important part of the appeal of canal travel.

This was one part of Rolt's writing, the recording of a way of life that would not last for ever. He was a passionate believer in the values of the canals, and where earlier writers had been content simply to view the picturesque sections and turn away from the rest, he celebrated them in all their diversity. He had a clear-eyed view of the threats that they faced from road and rail transport and was passionate in his defence of this extraordinary system. I can do no better than again quote his words from the final chapter of the book:

> In a society framed to cherish our national heritage the canals can play their part not only as a means of transport and employment, but as part of an efficient system of land drainage and a source of beauty and pleasure. But if the canals are left to the mercies of economists and scientific planners, before many years are past the last of them will become a weedy, stagnant ditch, and the bright boats will rot at the wharves, to live on only in old men's memories.

Narrow Boat was far more than just a magnificent account of canal travel, it was also, implicitly, a call to action. Rolt's vivid descriptions are matched by the two thoughts – this is a world under threat and it is one worth saving. The question was – what could be done about it? At the time, the answer was nothing at all, for the country was still at war. And even when peace arrived, the Government had rather more pressing problems to worry about. There were, however, people who read and responded to the book and decided that something needed to be done and decided that they were the ones to do it.

12

A PERIOD OF CHANGE

PARADOXICALLY, IT was war that, temporarily at least, prevented the inexorable decline in canal traffic that Rolt had foreseen. Every transport system was stretched to near breaking point and the canals had to play their part under ever-worsening conditions. Manpower was scarce, as the young went off to fight. There were neither funds nor the materials to carry out anything but the most basic maintenance and passage down some canals was almost brought to a halt. Silting reduced the channel, old gates stuck and leaked, paddle gear was left ungreased. It all made work twice as difficult, but there was one compensation; the canals were now under Government control and for the first time ever the boaters got one week's paid holiday. Few of them went off anywhere. As one pointed out, when you spend your whole life on the move, sitting still in one place is the best possible break. Canal work was also made a reserved occupation, but not before many had already left for the forces. The increased work load for those that remained began to strain the system. There was a Government drive to recruit crews, and far and away the most famous of the volunteers were young women, who were dubbed the 'Idle Women' – not because they were idle, far from it, but from their badge of office with the initials IW, for Inland Waterways. There has been a lot of attention paid to this labour force, largely because several of the volunteers turned their experiences into books. In reality their contribution was quite small, in that there were never very many at work. Several of those who joined up failed to survive the initial six-week training period, and many more never lasted beyond their first trip. Those who stayed worked hard. It was a matter of pride to try and equal the efforts of those who had been born to canal work and for whom the complexities of boat handling were virtually automatic.

The women who volunteered often did so because they had enjoyed boating as children, and it came as a rude shock to find the gulf between handling a fully-loaded 70ft boat and a yacht on the Broads. It was just as big a shock to the traditional boating community when the women arrived, as Emma Smith wrote in her book *Maiden's Trip*:

It must have been an astonishing imposition for the canal people when war brought them dainty young girls to help them mind their business, clean young eager creatures with voices so pitched as to be almost impossible to understand. It must have been amazing, more especially since the war changed their own lives so little, for they read no newspapers, being unable to read, and, if they did possess a wireless seldom listened to the news. For years, for generations, they had worked out their hard lives undisturbed, almost unnoticed. The suddenly – the war; and with it descended on them these fifteen or so flighty young savages, crying out for windlasses, decked up in all manner of extraordinary clothes that were meant to indicate the

Bomb damage at
Banbury locks on the
Oxford, 1942.

Volunteers on a Grand
Union Canal Carrying
Company boat during
the war.

Instructor Kit Gayford,
on the left with her
usual ear-flapped hat,
and a group of trainee
boatwomen during the
Second World War.

Volunteers posing on a GUCCC boat.

marriage of hard work with romance. For the most part the boaters took it stoically. They
watched narrowly, in silence, and they spat and they waited.

The women started their working lives under the eyes of two trainers, Kit Gayford and Daphne
French, the former of whom wrote a very matter-of-fact memoir, *The Amateur Boatwomen*.
The trainees – and the name stuck even after they had been on the canal working full time for
months – were not expected to help out with existing crews. Instead they made up their own
all-women crews, assigned to a pair of boats, mostly working on the Grand Union, but with
occasional forays onto the Oxford. They were resented by many of the boating community,
and had to work hard to earn their respect. Susan Woolfitt in *Idle Women* described a typical
conflict between themselves and a pair of 'jossers'. The women were about to set off up Hatton
locks and the first lock was almost ready, when the other pair appeared and headed straight
for the gates. Susan Woolfitt promptly started her engine and the two motor boats appeared
at the foot of the locks at the same time and jammed together. There then followed a slanging
match where, 'We told each other the most crushing truths we could think of, touched on
the doubtful domestic habits of one another's forebears, the state of the boats and the rights of
the case. They sneered at us and called us so-and-so and so-and-so trainees'. Eventually, with
a good deal of ill will, the women gave way and had to follow up Hatton with every lock set
against them. But they had sweet revenge later when they met the same boats firmly fixed in
the mud, a great embarrassment to any professional boater. The jossers had no option; they had
to be pulled off by the 'so-and-so trainees'. It marked a new relationship, and they became great
friends after that. Susan Woolfitt makes the point that if the women had simply backed down,
they would have been despised. But they had been firm and best of all they had helped out
,in spite of everything that had happened before. Many of the boating families overcame their
original mistrust and admitted that the trainees worked as hard as they did – and a good deal
harder than any of the men who had been recruited to the waterways.

The accounts by the women are all worth reading, for they are among the very few descriptions of what life was really like for those who worked on the canals. The trainees were the first to admit that they made more mistakes than the professionals, but it was generally the mistakes that make for the more entertaining reading. Margaret Cornish in *Troubled Waters* recounts many of them, including an alarming incident when the tow rope from the motor broke, unnoticed by the steerer, leaving the butty stranded in the dark in Braunston tunnel – a daunting 1¼ mile long. She also tells of the horrors of boating in an icy winter. The only way they could make progress was to unhitch the butty, then drive the motor at the ice at full speed, create a channel and then go back and hitch up again. The ice had to be broken up in the locks before the gates could be opened. In the end they were stuck for a week with a dwindling supply of coal to keep the cold at bay. It was an experience that was all too common to the boating community and her vivid description helps to give an insight into the 'romantic life' of the canals.

There was one aspect of life that was peculiar to the times. Out on their runs, the boats were comparatively safe, but when tied up at the docks there was the constant threat of bombs. Limehouse Dock was the worst place, particularly in the last months of the war when the V1 and V2 rockets were attacking London. Susan Woolfitt was in the cabin when a rocket hit the factory by the wharf and the blast caused havoc among the boats. In the end no one suffered more than cuts and bruises, though the boats were badly damaged. When the boats were patched up they were sent off again, far from spick and span. Other boaters were rather scornful of their condition and seemed to think it was no more than could be expected from women trainees, who didn't know how to look after things properly. Attitudes soon changed when the news was relayed with a certain pride – 'We've met a ROCKET'. It says a great deal for these women that the experience did nothing to deter them.

Then the war came to an end and it was very clear that the trainees were no longer needed. One by one they left the boats behind, often reluctantly – all but one. She married a boatman and became Mrs George Smith. It did not last – though they remained the best of friends – and Sonia remarried, this time to the man who had written so eloquently about the working life of the canals. She became Sonia Rolt and she has never lost her love of the canals. But before they were married, Tom Rolt was to become actively involved in a campaign to save the waterways for posterity. The first tentative steps were taken by a literary agent, Robert Aickman. He wrote to Rolt suggesting setting up a campaigning organisation for the canals: 'A sort of cross between the Light Transport League and the Friends of Canterbury Cathedral.' A meeting was set up at Aickman's London home, to which interested parties were invited. Rolt was not the only author to attend the first meeting – he was joined by Charles Hadfield.

Charles first became interested in canals as a schoolboy in Devon, when he used to walk along the towpath of the Grand Western Canal, which was already falling into disuse. My 1928 edition of De Salis records that at that time there were only two boats on the whole canal, chained together and carrying stone for local roads. But the boy was sufficiently interested to start asking about its history, and a local solicitor let him look through the surviving documents. It was the start of a lifelong passion, not just for canals but for research into old documents to tease out the facts. After graduating from Oxford with an Economics degree, he worked briefly – and unprofitably – in the second-hand book business, before joining the London office of Oxford University Press. He became interested in politics, and at the age of twenty-five he was elected as a Labour Councillor for one of the Paddington wards. He spent the war years as an auxiliary fireman in the river branch of the London Fire Brigade. He kept up his political activities and was soon a leading member of the Fire Brigades Union, and started up the country's first local union journal, the *River Service Bulletin*. Those of us who only knew Charles in later years find it rather difficult to think of that quiet and invariably courteous man as any sort of political firebrand. On the other hand, on the few occasions when we disagreed on canal matters, he was always firm and persuasive: one can easily imagine him as a formidable adversary in his union career. In 1942 he was seconded off to write the official *Manual of Firemanship*, as one of a team of three authors. One of his colleagues was Frank Eyre, who shared his enthusiasm for canals. Together they wrote *English Rivers and Canals*, published in 1945. It was his first canal book, but was to be followed by the hugely successful *British Canals*, an illustrated history, which has now been in print for half a century. After that he began work on the series of regional histories that have been invaluable guides for anyone with an interest in canal history ever since.

The Hadfield and Rolt were complementary figures, each providing a particular perspective on canals, but both sharing an enthusiasm and a desire to preserve what remained. They were both initially keen on the idea of starting a pressure group that shared their notions. At the time of the first meeting, the proposed organisation did not even have a name, but one of those who arrived for the meeting knocked at Aickman's door and enquired if this was where the Inland Waterways Association (IWA) meeting was being held. The infant had a name, and now needed an organisation. Aickman was appointed chairman, Hadfield vice-chairman, Rolt secretary and Eyre treasurer. It marked the beginning of amateur involvement in a campaign for the restoration of canals.

The IWA was formed at a critical moment in canal history, when everything was about to change. The first post-war election saw a Labour government come to power, and promptly set about a major programme of nationalisation, that was to include taking railways and canals into public ownership. On 1 January 1948 nearly all the canal system was taken over by the newly formed British Transport Commission. The bits that were left out consisted of several derelict or near derelict canals, such as the Rochdale and the Glamorganshire and a number of river navigations, including the Norfolk Broads. More surprising omissions were the Bridgewater, Manchester Ship and Exeter Canals, which for some impenetrable bureaucratic reason were considered to be primarily docks. The other exception was the non-tidal Thames, which had long been mostly given over to pleasure boating and was more than ably managed by the Thames Conservancy. Even with all these exceptions, the new Commission found itself the owners, if not exactly the proud owners of 2,172 miles of waterway, much of which was in a poor state of repair and losing money.

Everyone had had plenty of time to think about what should happen to the canals. Nationalisation had first been discussed at a conference held in 1888, and the take over of major routes had been recommended by a Royal Commission of 1909. The most important feature of the legislation, when it finally arrived, was that for the first time, canals could be thought of as part of a general transport policy and not in isolation. It was soon obvious that there were priorities to be established. The most pressing was to catch up on the neglected work of maintenance, but no less pressing was the need to spend money where it was most likely to show a return. The most promising candidates were the broad waterways and a programme was set in hand improving both the navigations themselves and the boats using them. The Severn and the Trent were the first beneficiaries, and soon modern motorised barges were able to travel much further up the rivers. Now vessels up to 400 tons could reach Worcester and 300-ton barges could travel to Nottingham. The new authorities also took a close look at what sort of traffic the waterways should carry. They identified a clear case for developing waterways connected directly with major ports. The emphasis would be on loading the barges directly from ships and sending them on their way, cutting out one whole stage in the handling process. A long term aim was to make the sort of improvements that would allow coasters to travel far inland. It was

Limehouse Dock, London was a target for bombs during the war, but by the late 1940s business was flourishing.

all very sensible, in as far as it went, but left a big question unanswered. What was to happen to the rest of the system?

There were attempts to modernise traffic on the narrow canals. A new type of steel-hulled narrow boat was introduced. The old cloths were replaced by curved fibreglass covers, each 2ft long, which interlocked to cover the whole of the hold. Assembly took no more than two minutes, far faster than fastening up the traditional cloths. The covers were blue and the boats became known, not very imaginatively, as 'blue tops'. Crews soon found that the redesigned hull was far more difficult to manage than the traditional boat, and 'blue tops' soon became 'dustbins'. The authorities were naturally keen to present a modern image to the world, and a new house style of painting in blue and yellow was brought in. One executive referred to the colour scheme rather scornfully as 'the Billy Smart Circus' look. Whatever new style had been chosen, it would not have been welcomed by the boating community. The boats were their homes, and they were proud of them and their decoration. Now they were being told to get rid of their roses and castles. This created an uproar that even reached that most august of protest outlets – the letter pages of *The Times*. A compromise was reached – blue and yellow remained on the outside; roses and castles would be allowed inside cabins.

Responsibility for canals was handed to the Docks and Inland Waterways Executive branch of the British Transport Commission. They had their own magazine, *Lock and Quay*, and as early as 1949 an article noted a new phenomenon: 'A good deal of interest has been shown in pleasure cruising on the canals.' It was thought that, if encouraged, this could be a useful source of income, though there were problems. The article noted that some canals would not attract the pleasure boaters and some, which were particularly picturesque, were scarcely in a condition to accept any kind of boat. It was not very clear then what, if anything, should be done about this new sort of venture. For the time being it was ignored, while work went on to try and make the freight business profitable. The new campaigning organisation the IWA was equally unsure of just where the future lay.

The first attempt at preserving a canal was a comparatively modest affair. Shortly before nationalisation, the Stratford Canal, then still owned by the Great Western Railway, had been effectively closed. A former drawbridge at Lifford had collapsed and been replaced by a solid steel structure that sat low over the water. The IWA realised that the terms of the original Canal

Post-war improvements to the River Lea brought an increase in barge traffic.

Act still applied, which enshrined the right to navigation. A newly recruited IWA member, Lord Methuen, raised the matter in the House of Lords, and the GWR had little option. They indicated that if any vessel wanted to use the canal and gave warning of their intention, the bridge would be lifted. Tom Rolt duly gave note that he did intend to take *Cressy* along the canal. It was a difficult passage – the canal was in a wretched condition – but eventually the bridge was reached, where a GWR work party had jacked the bridge up, high enough to let *Cressy* scrape through. The IWA now had to persuade as many members as possible to apply on different dates to make the same journey. The plan was that eventually they would become such a nuisance that the GWR would decide it would be cheaper to replace the low bridge with another lift bridge. It was an encouraging start.

It soon emerged, however, that there were fundamental disagreements within the organisation about the direction in which it should be heading, and where its priorities should lie. One group, headed by Aickman, believed that the IWA should be lobbying and working towards the preservation of the entire system. The other, of which Rolt was a leading figure, thought that the emphasis should be on the canals that had a commercial future, and that they should be working in support of the dwindling numbers of commercial carriers. In the second of his autobiographical books, *Landscape with Canals*, Rolt set out his argument:

> What chiefly appealed to me about the canal system was its indigenous working life. On the narrow canals this meant the working narrow boats and their crews which are an essential part of them. These working boaters, so many of whom I knew and admired, unconsciously supplied that subtle traditional patina of constant use – the worn and dusty towpath, the polish that generations of 'uphill or downhill straps' had given to the bollards or grainy oak at the locks; it was an essential part of that blend of utility and beauty which used to compound the particular magic of canals. This was something that some members of the IWA could never fully appreciate.

A pair of British Waterways
boats photographed on the
Grand Union shortly after
nationalisation.

Rolt argued that, given the limited resources available, the IWA should concentrate on the
canals such as the Trent & Mersey and the Oxford, which were still viable, and forget about
what he called the 'fringe waterways' at least for the time being. With hindsight it is easy to see
that narrow boat carrying was doomed and nothing that an organisation of amateurs could do
or say could save it. On the other hand, many would agree with him that it is the very nature of
the working canal and its long history of use, that gives it a unique character. The future may be
seen to lie with pleasure boating rather than commercial carrying, but if the sense of history is
lost, a canal becomes no more than an elongated boating lake. There was definitely a case to be
argued. Unfortunately rational argument lost out to a clash of personalities. Charles Hadfield
was the first to go. He had taken a new job at the Central Office of Information, which meant
that he could no longer be an officer of a campaigning organisation. Rather more importantly,
he did not get on with Aickman. Ironically, in view of later events, it was Rolt who came to
Aickman's defence:

> Aickman has not, I know, an easily likeable personality, and is apt, on this account to antago-
> nise people unreasonably. But he is remarkably enthusiastic and efficient, he has devoted an
> immense amount of time and labour to the IWA, and without his efforts it is extremely
> doubtful whether the Association would ever have been born at all.

Rolt had also embarked on a new career as a professional writer, which was demanding
more and more of his attention. At the same time, the IWA seemed to think that he and
Cressy should be at their disposal whenever it was thought they would help to attract publicity. Things came to a head when Rolt received a letter telling him to attend a meeting
at Newbury in three days time 'couched in terms which I would have hesitated to use to
anyone'. He did not go, and resigned his position as Secretary on the grounds of new work
obligations.

It seemed a perfectly valid reason for resigning, but it unleashed a furious response. Rolt was more or less denounced as a traitor. The very lowest point arrived in 1949. Rolt had suggested a boat rally at Market Harborough, which would publicise the IWA and also be a chance to gather supporters together from across the country. The idea was approved, but before the date arrived, Rolt had sent in his resignation. He then received an official letter saying that he and *Cressy* would not be welcomed at the rally. It was an extraordinary letter to write. Rolt had suggested the rally in the first place; *Narrow Boat* had been the inspiration for many to take an active interest in canals, and he and *Cressy* would have been the main attraction for many who came to the rally. Apart from the fact that there was no way Rolt could be prevented from travelling on a public waterway, it seemed both churlish and inept. The inevitable result was that Rolt resigned from the IWA and was soon devoting his time, and not inconsiderable energies, to another love – railway restoration. If the object in banning Rolt had been to avoid making him the focus of opposition to the official view of the Aickman group, then it had quite the opposite effect. When Rolt left others followed. If, on the other hand, it was all part of a plan to remove all likely opposition to the official line, then it was entirely successful. The main participants in the crisis are all dead now, but the bad feeling never went away. When Robert Aickman died, I asked Charles Hadfield if he would be going to the memorial service. He merely replied that if he turned up, Aickman would rise from his grave in horror. I left it at that.

In the immediate post-war years the whole future of the canal system was in the balance. There was a general agreement that it was worthwhile improving some of the broad waterways, but no one was altogether certain about what to do with the rest. While the whole matter was being debated, the position was completely altered by a change of Government. The new Conservative administration largely abandoned the idea of a unified transport policy for the whole country. Out went the Docks and Inland Waterways Executive and in came a new Board of Management, with somewhat reduced powers. The emphasis was now to be all centred on profitability, but first someone had to find out the basic facts and produce a profit and loss account. When the work was done it did not make cheerful reading. The report of 1953 showed that less than half of the waterways showed a small profit; the rest showed a loss. The proposal set out what should be done, which involved splitting up all the waterways into a league table. The main emphasis was on the Group I waterways, the 336 miles that showed the best return, and which should be improved. Then came the Group II, which were of dubious value, but which it would be worth keeping going with a minimum of maintenance and efforts should be made to attract more traffic. If the traffic did not materialise, then they would be relegated to Group III. That accounted for another 994 miles. Group III itself was a mixture of canals that had either fallen into disuse or were very likely to do so. Here nothing should be done; they could either be left to decay, or handed over to anyone else who wanted them.

There were some big names in Group III, including the Kennet & Avon, the Llangollen and the southern end of the Oxford. The future looked gloomy for some of Britain's best loved waterways. The proposals made commercial sense in terms of freight carriage, and if nothing else had been taken into account, one could see what would happen. The Group I waterways would have a future, but many of the Group II would fall into the black hole of Group III, never to emerge again. If the goals of the IWA were to be realised, and the network saved, then there would have to be new and compelling arguments for keeping canals open. The statistics made it all too clear that freight was very unlikely to play a major role on the narrow canals, so there had to be an alternative traffic to justify their existence. There was. The old working canals would be redeveloped for leisure boating.

13

RESTORATION

THE NOTION that canals could be revived and could thrive on the basis of holiday traffic and nothing else, was not an idea that was greeted with any great enthusiasm by the authorities. The IWA approached the British Tourist and Holiday Board in 1947, a body they must have thought of as natural allies. They were given a frigid reception: 'Regarding the development of rivers and canals, I have been in communication, through the Board of Trade, with the Ministry of Transport. The latter do not look very favourably upon any scheme for pleasure craft on the canals at the present time.'

It was clear that if anything was to be done to preserve and restore canals for leisure use, it was not going to be the Government that led the way. The first opportunity for positive action came in 1949 with the announcement that the Basingstoke Canal was to be put up for sale. It had been bought by a private company in 1923, but the owners had totally failed to make it a commercial success. Public opinion was divided; some were happy to see it abandoned and filled in. One has to remember that at this time, canal pleasure boating was an activity enjoyed by a very small, almost minuscule, proportion of the population. To very many, canals were simply nasty dirty things, home to dead dogs and potential death traps for small children. There was, however, a significant local movement in favour of the canal, and the IWA decided to campaign for keeping it open. The organisation had no money to buy it, but they were enthusiastic and persuasive advocates, and publicity brought in the necessary donations. In the end the canal was purchased on behalf of the newly formed Basingstoke Canal Committee of the IWA by a Mrs Greenways, who lived near the canal at Fleet. A company was formed to run it, the New Basingstoke Canal Company. The hope was that revenue could be earned, mainly from fishing licences and residential moorings. The IWA were delighted with the result, but it soon became apparent that the optimism was misplaced. The revenue never materialised and the canal began a slow slide into dereliction. It was to be many years before a practical rescue plan was put into operation.

The 1928 Bradshaw had actually made the point that the Basingstoke was an ideal canal for pleasure boating, particularly the 21-mile pound from Basingstoke to Ash Lock, which was described as passing through 'unsurpassed rural scenery'. This was very true, but the canal suffered from being somewhat isolated from the rest of the system. The eastern end was approached from the Thames via the River Wey, but in the west it simply came to a dead end at Basingstoke. Another canal had rather better connections.

The Kennet & Avon is linked to the Thames by the Kennet Navigation and to Bristol and the Severn by the Avon. It too was under severe threat and strenuous efforts were made to keep it open. The canal was one of many that had been taken over by railway companies, in this case the GWR who acquired it in 1852. As one of the conditions they were required to keep it open, a point that was emphasised by the 1873 Regulation of Railways Act. It was one thing to keep it open; quite another to encourage its use. It was rather as if they had read Arthur Hugh Clough's modern version of the Ten Commandments:

Thou shalt not kill; but need'st not strive
Officiously to keep alive.

As a result, by the time the British Transport Commission took it over, the canal was in a wretched state. They showed as much enthusiasm for improvement as the GWR had done. If the canal was to be saved, then action was needed. The Basingstoke model was followed and a new branch of the IWA created in 1949 to fight the threat of closure. Newbury was one of the keys to the retention of the system; it was here that the canal met the older River Navigation. To the dismay of the enthusiasts, the local council decided that they were opposed to any attempts at restoration. One local resident decided to take the fight to the council. John Gould knew the waterway well and in 1949 he bought two narrow boats and announced that he was going to bring trade back to the Kennet & Avon. He arrived in Newbury with his first load, paving stones from Birmingham, in November of that year. The authorities were not pleased: officialdom wanted to see the end of the canal not its revival. In 1950, all the locks and swing bridges were closed, with no previous warning. It seemed as if permanent closure was inevitable. One fight might have been lost but the battle was far from over.

A decisive moment came when the local enthusiasts split from the IWA to form their own campaigning organisation, which eventually became the Kennet and Avon Canal Trust. John Gould started a petition and collected 22,000 signatures to oppose the proposed Act of Closure. With such a huge level of support, the Act was withdrawn before it ever reached Parliament. Officially, at least, the canal remained open; in practice it was totally un-navigable. In those early years of restoration movements, organisations relied on amateur fundraising and working parties. The funds were limited, and so was the amount of work that could be done on a part-time basis. Those of us who were involved in the early days on the Kennet & Avon probably all have memories of clearing a small section of canal and moving on. But because there was no regular maintenance, the first section would be starting to deteriorate again, before the next one was finished. There were real successes, however, to set against this seemingly never ending fight against decay.

The restoration of the two pumping stations, the steam powered pumps of Crofton and the water-powered site at Claverton, was an immense achievement. These are sites of national, indeed international, importance. I cannot think of anywhere I have visited that more dramatically illustrates the power and importance of the steam engines of the pioneering firm of Boulton & Watt than Crofton. There are bigger engines elsewhere, but none with quite the same atmosphere – and none where you can see so clearly exactly what they do. On steam days, the original Boulton & Watt engine can still do the job it was installed to do, pumping water up from a lake and into the summit level of the canal. Working with the second engine, the pair of them can shift a quarter of a million gallons an hour, and at each stroke of the pump another gush of water can be seen roaring out into the leat. If working steam engines are rare, working water-powered pumps are even rarer, and among the few survivors Claverton is a giant. It has two coupled wheels, each 15ft 6in diameter and 11ft 6in wide, and although they may not match the steam engines for power, the pumps can still lift 100,000 gallons an hour to the canal 47ft higher up the hill.

A peaceful scene on the Kennet & Avon
near Claverton, a restoration success story.

Crofton and Claverton were triumphs, and the complete restoration
programme was eventually a triumph
as well. In 1990, John Gould shook the
hand of the Queen, when she came to
the official re-opening, and no one had
done more to earn the privilege. But the
success was not entirely due to the hard
labour of occasional volunteers. The
Canal Society became a Trust, a fundraising body, using the money to pay for
professionals to take over the main body
of the work. It has to be said that if it
had been left to the piecemeal efforts
of the amateur labourers we would not
be seeing boats making the trip from
Bristol to Reading today. Other restorers
had already formed the same judgement:
do it yourself was not the only route to
canal restoration.

A start had been made at keeping the
Stratford Canal open and enthusiasts
began a new dream. Wouldn't it be wonderful if restoration of the canal could be accompanied
by the reopening of navigation on the River Avon. Then pleasure boaters would be able to
enjoy a holiday going on a circular trip, instead of just going up the canal and back down again.
They would do the Avon Ring. The prospects were not good. There was some traffic on the
Lower Avon in the 1920s, when pleasure steamers were run at Evesham and Tewkesbury, but
the steamers had gone and traffic had tailed off almost to vanishing point in the post-war years.
The Upper Avon was in a far more parlous state. The last boat had used it in 1873.

The Lower Avon had not been included in the nationalisation package and the Lower Avon
Navigation Company was only too pleased to get rid of it. The Midlands branch of the IWA
offered to buy it, and the price is a good measure of the enthusiasm with which the owners
regarded the waterway. The agreed price was £1,500. It is one thing to use volunteers to work
on a canal owned by somebody else, as on the Kennet & Avon, but it is a very different matter
to take over the entire responsibility for both restoring and later maintaining a waterway. The
Lower Avon Navigation Trust was formed and began to raise the money for restoration. The
work was not entirely left to the amateurs. The Royal Engineers stepped in and offered to
help rebuild Chadbury Lock, which was a good start. There were some interesting problems
to overcome. So little had been spent on modernisation, that two of the locks were still the
old-fashioned flash locks, among the very few survivors to be found anywhere in Britain. If
they had been retained it would certainly have made a trip on the Avon an unusually exciting
experience. Visitors to the present lock at Fladbury might take a look at the weir by the mill
and contemplate charging down something like that in their boat, for this was the site of one
of the navigation weirs. Given the problems, it is scarcely surprising that the navigation was not
open until 1962.

Restoration work is by no means an all male activity: a volunteer on the Thames & Severn.

The Queen Mother at the reopening ceremony for the Stratford Canal.

Before that happened, the emphasis had shifted to the southern end of the Stratford Canal. The northern end had been an important part of the canal network and well used, but the southern had fallen into decay. Rolt had managed his famous journey to the lift bridge, but his had been the last narrow boat to use the canal. In 1959 Warwickshire County Council applied for permission to abandon the waterway, quoting the law that allowed them to do so, provided no boat had made a passage in the previous three years. But one had. Not a narrow boat, but an enterprising canoeist had made the trip and had kept all the toll receipts. That was all the evidence that was needed to prevent legal closure. The IWA had a powerful ally when they campaigned for the Stratford. The National Trust indicated their support and in the event they were able to secure a five-year lease on the canal. Restoration could begin, and this time circumstances were more promising. The total cost was estimated at £35,000 of which the Government offered to pay £20,000 and the Pilgrim Trust came up with a very welcome £10,000. Work could start on the first restoration of a derelict narrow canal, and the work was put under the direction of David Hutchings.

Anyone who has ever met David Hutchings has come away with the impression of a man fizzing with energy. He was not going to wait for volunteers to come forward. He went out to look for a workforce of any kind, from any source. He found just what he wanted at Winson Green Prison in the suburbs of Birmingham. He persuaded the authorities that doing positive work for the community and learning new skills in the process would be a major step towards prisoner rehabilitation – as well as a source of free labour. The prisoners who came to work on the canal did so as volunteers. Other workers were more likely to have been volunteered, as men were brought in from both the army and air force to help with the task. Restoring the canal might have seemed quite enough for one man, but Hutchings also took on the equally daunting task of restoring the Upper Avon.

There are two possible approaches to restoration – one is to make the preservation of the historic fabric a primary objective; the other is to push ahead to get navigation open as quickly and as efficiently as possible. David Hutchings was very much an enthusiast for the latter approach. It was not that he was unaware of history, but it was never the priority. He was also not afraid to take bold decisions. When a new lock cottage was needed at Evesham it would have been simple and uncontroversial to have designed a conventional building with brick walls and a pitched roof of slate or tile. Instead an uncompromisingly modern A-frame building was constructed. There were a few protests, but the engineers and architects of the twentieth century were only doing what their predecessors of the eighteenth century had done – they built using the best available materials and the latest ideas in design. The opening of the whole of the Avon ring was a big success for the restoration movement, but some elements were never to be repeated. Servicemen and prisoners were no longer available as a workforce. It looked as if it was back to the volunteers, and the small, ad hoc weekend working parties. Many enthusiasts, however, felt that the process was simply too slow for any meaningful results. It was time for some new thinking.

Anyone who has been on a canal work party will know that the volunteers can be roughly divided into two groups: those who put up with the hard work and the muck because they believe in the end result and want to do their bit; and those who revel in the job itself and are never happier than when up to their knees in mud. I confess to being one of the former. A member of the London and Home Counties branch of the IWA was one of the latter. He started a new magazine, *Navvies Notebook*, in 1966, which had a special appeal to the mudlarks. His name was Graham Palmer. In many ways he could not have been less like David Hutchings. When Mikron Theatre Company staged a show 'Mud in Your Eye' based on canal restoration, they were characterised as Superhutch and Garden Gnome. It was easy to see where the names originated.

Hutchings was focussed and almost puritanical in his dedication to the job. Graham Palmer, with beard and woolly hat certainly had a gnomic appearance, but in his own way he was just as driven. But for Graham the work had to be fun and the pints that flowed at the end of a working day were as much a part of the canal scene as shovelling dirt or bricklaying. Graham recruited a small army of like-minded volunteers to form the Waterways Recovery Group, the WRGs.

The idea behind WRG was simple. Getting a large group of people together in one place for a short period of concentrated work is more efficient than having small groups scattered all over the place over a long period of time. The theory was soon put to the test. The success of the Avon Ring suggested that an equally successful circuit could be developed further north, the Cheshire Ring. The two sections that were going to cause trouble were the Peak Forest and Ashton Canals. The Peak Forest had a more obvious appeal, and included one of the system's grander aqueducts at Marple. The Ashton, it has to be said, is less immediately attractive. Seventeen locks are crammed into just under 7 miles of canal, which lead through the old, but sadly run-down industrial outskirts of Manchester to the heart of the city. It was easy to sell the idea of restoring the Avon and the Stratford Canal, which were undeniably picturesque. If the Ashton was ever to appear in a picture, then it would be in one of Lowry's gloomier paintings. Indeed, looking at a Lowry would give a fair notion of what the Ashton was like in the late 1950s. Nevertheless, there were many who thought it worth saving, not just for the sake of the Ring, but because it is a reminder of what the canals were all about – serving the Industrial Revolution.

By 1959 the British Transport Commission had given way to the British Waterways Board, BWB, who announced that while the Peak Forest was still officially open, it was more or less un-navigable. The IWA decided to check this by organising a small flotilla to cruise the Ashton and Peak Forest. The night before the boats set out, one of the locks burned down; whether this was an unfortunate coincidence or a deliberate act by someone who hated the whole idea of reopening the canals was never discovered. It was not a promising start. Worse news followed in the harsh winters of 1962 and 1963, when frost caused severe damage, particularly to Marple aqueduct. Conditions on both canals deteriorated.

The Ashton suffered from the fate of many neglected urban canals. Without a regular passage of boats, it became the dumping ground for any old rubbish from bedsteads to the ubiquitous supermarket trolley. Clearing it would be a massive operation, but that was the task the WRGs took on with military precision in Operation Ashton. Six hundred amateur navvies descended on the canal on just one weekend in 1968. By the end, vast piles of rubbish had been burned and what couldn't be incinerated was carried off to the dump – 2,000 tons of it. It showed everyone just what could be done. Eventually the Peak Forest and Ashton Canals were reopened, which involved major work on the locks on both waterways. I first travelled them by boat in the 1970s, and it was instructive to see different approaches to restoration problems.

Descending the Marple locks, it was obvious that the restorers had begged, borrowed or somehow acquired a bewildering mixture of paddle gear. Some worked perfectly; some could just about be persuaded to move. It made for an interesting, if frustrating journey, not improved by the fact that it rained relentlessly all day. Yet it was a rewarding experience, for this is a flight of locks full of interest. The locks have side ponds, which serve as unofficial garden pools to the householders lucky enough to live by the canal. But it is the canal itself and its structures that supply the main interest. It passes under a busy main road, and the towpath goes through a short tunnel, with an appropriately horseshoe shaped profile. At one end, it emerges by a flight of well-worn stone steps leading down into a typically, deep, dark and forbidding lock. Further down the flight is a reminder that this, too, was a canal serving industry. A warehouse, with an arch over the canal for loading, was built by the cotton magnate Samuel Oldknow, who was

The Rolle aqueduct – now a driveway – is one of the few surviving structures on the Torrington Canal.

one of the main promoters of the canal. What really gives the whole canal its sense of cohesion and unity with the surrounding countryside, is the use of local stone, not genteelly carved, but left in great rough blocks, in buildings and walls. This is a restoration that, whatever problems may have been faced, has left us with a canal of real character.

The Ashton inevitably proved very different. At first sight it seemed that here, too, the old had been respected. The waterways authorities and local industries had got together to make the canal a real focal point for the area, something which could be enjoyed as much by locals as by boaters passing through. Balance beams were crisply painted, the lock surrounds had been given chunky, stone setts, with ribs for providing a grip when opening a lock gate. The idea is that you lean your back into the balance beam and walk steadily backwards to open the lock. Now the lock scene looked wonderful, but at the first lock I began opening the gate, only to find that one of the newly painted bollards was right in the middle of the walkway, which is not the best place to find one when you are walking backwards. Handrails were, irritatingly, only supplied over one set of gates. In one thing, at least, I could be called a Luddite. I do not like hydraulic gear. It is allegedly more efficient, but not in practice, and it has none of the quirky charm of the old spiky, mechanical gear. I had no doubt at the end of this part of my trip which restoration scheme got my vote – I preferred the rough and ready virtues of the Peak Forest to the urbane Ashton.

The Ashton marked the start of a new campaign, which saw ever bigger work camps set up on restoration sites. Among the canals that benefited was the Basingstoke, where one of the major efforts was concentrated at Deepcut, which is a length of the canal that is exactly what its name suggests. At places the cutting is 70ft below the surrounding area, and it ends in a flight of fourteen locks in 2 miles. It is amazing to see how much silt can accumulate once a canal is abandoned. The bottom of the locks was a smelly, black, oozing morass of heaven knows what, all of which had to be cleared. Much of it was done by hand, shovelling the muck into barrows,

The Rochdale Canal was privately owned and this section in Manchester remained open after the rest of the canal fell into disuse.

The Rochdale Canal, out in the Pennines, as it appeared before restoration.

then wheeling them away over planks all but indistinguishable from the surrounding muck. Deepcut was a WRG delight.

Perhaps the most remarkable restoration story, involving massed action by volunteers took place in the unlikeliest of situations – the heart of the Industrial Midlands. The Dudley Number One Canal is just over 4 miles long and forms a connection between the Birmingham main line and the Stourbridge Canal. In strictly practical terms there would be no case for keeping it open, since the same connection is made by a parallel canal, the Dudley Number Two, which offers a far easier passage. But the original canal has a unique feature, the 3,172 yard-long Dudley tunnel. This is no ordinary tunnel, but a route that leads to an underground labyrinth of limestone mines and quarries. In places it is lined, but elsewhere it is open to the bare rocks. In some sections boats squeeze through openings like fingers sliding into a glove; in others it opens out into soaring arches like the crypt of some grand church. It is, quite simply, unique.

In the 1950s traffic was negligible and in 1959 the BTC called for closure. But by then an unusual mix of cavers and canal enthusiasts had discovered the delights and fascinations of this underground world. A mass protest cruise was organised by the IWA, which produced a stay of execution, but it was no more than a short reprieve. In 1962 the tunnel was officially sealed and navigation prohibited. At that time it could still have been reopened if officialdom had approved, but in 1963 a far more serious threat emerged. British Railways applied to replace a viaduct by the north portal with an embankment that would be built right over the canal. There was an immediate protest, and in 1964 the Dudley Canal Preservation Society was formed. The threat was still being resisted when the railway authorities themselves decided that the Dudley line wasn't very important anyway and actually lifted one of the two tracks. There was no need for an embankment.

The movement for restoration received a huge boost, when it was decided to establish the new Black Country Museum at the northern end of the tunnel. It was now clear that there would be a real possibility of running trip boats through the tunnel as an extra attraction for visitors. Restoration was officially approved. As with many schemes, the early years saw work going on slowly in a piecemeal fashion, but in 1970 a 'Dudley Dig and Cruise' weekend was announced. It attracted fifty boats and 600 amateur navvies. There was a second major work party in 1971, and by now earth moving equipment had been added to old-fashioned muscle power. In 1973 the canal and tunnel were back in business. Since then the museum has grown to become one of Britain's most important open-air museums, and the tunnel trips have proved an immense success. If it does nothing else, a trip provides an opportunity for passengers to have a go at legging through a tunnel, if only for a short way. But even today, only a part of the underground complex is open. There is an old story that if you explore one of the many side tunnels leading to old mine workings and make your way to the surface you will emerge among the bears of Dudley Zoo. I have never tested the story.

The growth of pleasure traffic on the canals and the success of the early restoration schemes began to change public opinion. Two hundred years after the Trent & Mersey pamphleteers had extolled the virtues of 'a lawn terminated by water', house owners began to wake up to the fact that a canal was not something to hide away behind a high fence, but was actually an asset to be enjoyed. Even more importantly to many – it could actually increase the value of a property. Local authorities began to take a more favourable view and instead of arguing for closure began to think how a canal could be a real amenity to be enjoyed by everyone. There was still a long way to go, but a real change came with the economic depression of the early 1980s. As unemployment rose, so the Government began a search for jobs to give the young at least an opportunity to gain some sort of work experience, rather then going straight from school to the dole queue. Job creation schemes were run by the Manpower Services Commission,

The author, invited to officially
mark the start of restoration on the
Huddersfield Canal, demonstrating
his ineptitude with a pneumatic drill.

A rural section of the restored
Huddersfield Canal near Marsden.

and canal restoration was among the beneficiaries. But the big change came when new funding became available, especially through funds from Europe and the Heritage Lottery. Now schemes that seemed unimaginable just a few years before were suddenly not just under consideration, but activated.

Back in 1974, I had taken a trip to look at what remained of the two Pennine routes that had been abandoned – the Rochdale and the Huddersfield Narrow. What I saw impressed me, but at the time I wrote I could see no prospect of anything been done, and if I had been taking bets I would have said that neither would ever be reopened in my lifetime. The obstacles seemed too formidable. I was not only wrong, but I was actually given the honour of taking part in the opening ceremony for restoration of the Rochdale. It still seemed an impossible task. In 1987 I had walked the Rochdale from end to end, and apart from the immensity of restoring a canal with ninety-two locks in 33 miles, there were two seemingly impassable obstacles. At Farnworth, the Co-op had built a supermarket on top of the original line of the canal and the M62 motorway cut right across it on an embankment. But everything it seems has a solution, if you have enough money to throw at it. New land was bought, and a new supermarket built for the Co-op, allowing the old to be removed. The motorway embankment was pierced by a tunnel. In 2002 the canal was reopened.

The Huddersfield seemed even more daunting. This was a canal I have known for many years, as Mike Lucas of Mikron Theatre lives in one of the old lock cottages, and we often strolled down the towpath along the derelict canal towards the greatest obstacle to progress: Standedge tunnel, the longest canal tunnel in Britain, long since abandoned. Once again obstacles were overcome. In Slaithwaite the canal had long been buried, and many locals would have been happy to see it left that way. But a new channel was cut, lined with sheet piling and concrete, which was later given a respectable covering of stone, and today it seems as if the canal had never gone away. With problems such as these the fact that two factories had been built over the canal seemed quite minor problems. Restoration has been completed and Mike Lucas thought he would finally achieve his dream of mooring up the theatre company's narrow boat *Tyseley* outside his own house. With the reopening, in 2001, it seemed the day had arrived and a show tour was arranged along the canal. But disaster struck. *Tyseley* reached one lock, entered and stuck. Nothing it seemed would move her. Eventually the boat was dragged back out, but there was no way forward. Mike's house was never reached.

The most spectacular restoration project of all involved the reopening of the Forth & Clyde and Union Canals, to form a through route from Glasgow to Edinburgh. There were many obstacles to overcome, much as there had been on other canals, but there was one major difficulty. Originally the two canals had been connected via a flight of eleven locks at Falkirk, but they had long since disappeared. There were two options: rebuild the locks or replace them with some form of mechanical lift. It was decided to take the bold decision and go for a lift, and the result is a great work of modern engineering, the Falkirk Wheel. The idea of a lift based on counterbalanced caissons was not new; it is the obvious way of reducing the power needed for operation. In the past, the caissons were lifted in a vertical frame, but here they are set on opposite sides of rotating arms. Each caisson is held within a ring at one end of the arms, and is mounted on wheeled bogies. As the arms turn, the caisson automatically stays at the lowest point of the ring, so no water is spilled. After a rotation of the arms through 180 degrees, the caisson that was at the bottom will now be at the top and vice versa. It is not just brilliant engineering, but it is a remarkably beautiful structure, as shapely as a Henry Moore sculpture. There is an attractive footnote to this story. The steel for the construction was supplied by the Butterley works in Derbyshire, founded by two pioneers of the structural use of iron in canals, William Jessop and Benjamin Outram.

A major task facing volunteers of the Cotswold Canals Trust: Inglesham lock on the Thames & Severn.

Canal restoration continues, but the money is becoming increasingly difficult to find and the costs keep escalating. I live very close to the Stroudwater Canal which, together with the Thames & Severn, has been the subject of a long-running restoration programme. Back in the 1980s I was invited to present a television appeal to raise funds for the canal. We spent a weekend with a BBC crew, mostly filming in and around Sapperton tunnel and the result was shown on the Sunday evening public appeal programme, known in the world of television as the 'God Slot'. We raised, I believe, something in the region of £20,000 and that was considered a big success. Work is now under way on a 6-mile section, with a budget of £24 million. A quarter of a century is a very long time in the restoration world. The Cotswold Canal Trust remains determined to carry on until boats can once again travel through the Cotswolds from the Thames to the Severn. Having seen what has been done in other places there is no reason why they should not succeed. It just may not happen very quickly.

Only a few of the very many canal schemes spread around the country have been mentioned. They have shown many different approaches, but they have all had one thing in common. The initial enthusiasm and drive had all come from amateurs, ordinary members of the public who just happened to think Britain's canal heritage deserves to be preserved and was worth a fight to keep it. The work goes on, and covers an immense range of projects from the Bude Canal down in Cornwall, with its historic array of inclined planes, to restoration programmes for canals which, until recently, few even knew existed, such as the linked Lichfield and Hatherton Canals in the Midlands. Canal restoration is still a very long way from reaching the end of the line.

14

CANALS TODAY

THE MOST striking feature of the canals today is the almost complete absence of freight traffic on the narrow canals, yet the canals themselves have probably never been busier. In January 2009 British Waterways issued statistics that there were 30,000 boats on the canals and 11 million people a year were using them for leisure. Yet this leisure industry was still in its infancy half a century or so ago. Unlike the professional carriers, the small band of amateur boaters had very little idea what any particular canal would be like or what they might expect to find round the next bend. The De Salis Bradshaw had long been out of print, so in 1950 the then Secretary of the IWA, Lewis A. Edwards, produced his *Inland Waterways of Great Britain and Northern Ireland*. It listed all the navigable canals and provided a wealth of information on each one, including distance charts. Although it was aimed squarely at the holiday boater it is notable that the only illustration shows a working narrow boat. It was also thought necessary to put in a cautionary note from L.T.C. Rolt:

It is unfortunately true that the prejudice which some canal officials and canal boatmen entertain against pleasure craft is not altogether without foundation. In the past, certain owners of pleasure craft have tended to forget that the canal is first and foremost a commercial highway. Also that although his boat moves slowly, every minute counts to the working boatman, who is not paid by the day but by the ton carried.

The message that followed filled in the details of the basic rule: working boats should be given priority. Interestingly it was still thought necessary to remind boaters not to moor on the towpath side of canals used by horse-drawn boats.

In the early years, pleasure boating was more or less limited to people who owned their own boats, but gradually entrepreneurs began to realise that new commercial opportunities were opening up. As the freight traffic died away, so a few narrow boats were put to a new trade. They were given comfortable cabins and converted into hotel boats. There was also another new market appearing. Not everyone either wanted or could afford a boat of their own, but they were quite keen to try a canal holiday. Hire boats were already common on popular waterways such as the Thames and the Broads, and soon they began to appear on canals as well. This in turn created a demand for boats specially designed for this new trade. Among those who took advantage of the situation was George Wilson. He bought an old wooden Thames launch, restored it and took it for a trip on the Llangollen Canal in 1957. He loved the experience and was convinced that others would be equally enthusiastic. He decided to go into business as a boat builder and the result was the Dawncraft class of boats,

As narrow boat traffic died, some of the old boats were converted to other uses. *Tyseley* spent some time as a floating restaurant before becoming the travelling home for the Mikron Theatre Company.

which became the mainstay of many of the early hire fleets. They were still very much based on river craft design.

In 1956 a new man, Sir Reginald Kerr, took over what was now called British Transport Waterways, soon to be the BWB. He was convinced of the importance of the leisure industry for the survival of the main part of the canal system, and decided to set up a small fleet of hire boats. He also inaugurated a series of canal guides, each one covering an individual canal. They were simple in conception. The canal was represented by a linear map – no indication of twists and turns, but with all the bridges and locks clearly marked and a very useful distance scale running down the side. Alongside this was a simple descriptive text. They were cheap and they did the job. In those days 'canalling' – a term coined by Charles Hadfield – was very much a minority pastime and a lot of people coming to canals for the first time had very little idea what to expect. I was very much one of that number.

My wife and I had enjoyed a couple of canoeing holidays on rivers, when friends came back from a canal holiday and showed us photos, including shots of them crossing Pontcysyllte. It looked great fun and the next year we took ourselves off to the British Waterways hire base, on the Middlewich branch of the Shropshire Union, ready to head off to Llangollen. In those days the smallest craft available were known as Water Babies, little fibreglass cruisers, powered by an outboard motor. There was a small cabin just about big enough for the two of us. Armed with our guidebook we set off. In those days there was still some commercial traffic, and we had no need to listen to Tom Rolt's advice about who should have precedence. The narrow boat looked huge and terrifying – we headed for the bank and clung to it until the monster had gone past. We thought the tiny boat would be easy to handle. We had not realised that it was so light that it only needed someone to sneeze on one bank to send you skittering towards

The smallest of the new British Waterways Board's first hire fleet, a Water Baby; in the 1950s it cost as little as £10 a week.

The grander version of the Water Baby owes its style to the older river cruisers.

Hull is one of few places where working barges can still be seen in large numbers.

the other. Outboard motor and reedy canal rapidly proved an unhappy combination. It was to be the first of very many canal trips, but the last in such a vessel. We discovered a far more satisfactory alternative. A new type of craft had appeared, built by companies such as Harborough Marine, based very much on the traditional narrow boat, with steel hull and tiller steering. It made sense. The narrow boat had evolved specifically to meet the special needs of canal travel. It was robust and manageable and in time was to dominate most hire fleets and prove equally popular with private owners.

It is unfortunate that along with the development of the holiday narrow boat there also developed a form of canal snobbery. Anything that was not 'traditional' became a 'noddy boat' to be looked on with scorn. The people on board were generally included in the derision. There was a feeling among some traditionalists that they were the natural successors to the working boat families. There is only one qualification for being a working boatman – working. Stencilling roses and castles and buying painted water cans do not qualify. Whether we arrive on the canal in a converted narrow boat or a fibreglass cruiser we should admit we are there because of the pleasure it brings and not to earn a living. If we do our bit towards preserving traditions then that is all to the good, but we are doing it because we choose to do so. It does not give us the right to look down on others.

The increase in holiday traffic and private ownership brought a new problem to the canals. In the working days most boats were on the move for most of the time. Today many boats, especially those which are privately owned, spend a lot of time tied up to the bank. There was a danger that the canal system would become a linear parking lot. The solution was to get moorings off the navigation: provide marinas. There was a good deal of opposition to the whole idea from those who thought them out of character with the historic network. It is, however, difficult to see any satisfactory alternative. Systems need to adapt to changing circumstances, and for the narrow canals that means being fully occupied with busy holiday traffic. It is not the same for all Britain's waterways.

The broad waterways have a potential for increasing the amount of cargo they carry, which sadly the narrow canals do not have. The reason is simple. Much of the world's trade is carried in containers, and the smallest container currently in use is 8ft wide. An 8ft-wide container will not go though a 7ft-wide lock. No such problems appear on the wide canals and rivers. Here a long tradition of carrying has continued, with inevitable modernisation and updating. The experience of one particular boatman I know typifies the change. Geoff Wheat was one of those who carried on trading on the Leeds & Liverpool Canal long after others had given up. He had a traditional wide boat and was going along quite happily until the day when a crane driver dropped a load. It went straight through the hull and the boat was lost. But Geoff still carries, now as part of the Humber Barges group, based in Leeds. Where his old boat had a capacity of about 40 tons, the new motorised barge can take up to 480 tons. This is the equivalent of about twenty-four trucks on the road, and currently barges such as these carry the equivalent of 64,000 truckloads per year. It is Government policy to improve on these figures, but it has to be admitted that even so it is a very marginal effect in terms of freight transport over the whole country. To put it in perspective you need to look across the Channel to mainland Europe. The busy port of Rotterdam has links by road, rail and waterway and the latter are far from insignificant. The most important traffic is in containers and the container has provided a new measure of cargo carried, taking over from tonnage, the TEU, Twenty-foot Equivalent Unit. In other words total carrying figures are quoted in terms of 20ft-long containers. The last set of figures for barge traffic showed an astonishing 2.445 million TEU carried on barges.

Statistics are impressive, but why are these important? We live in an age where global warming threatens the entire planet with catastrophe and the biggest problem comes from carbon dioxide emissions. Right back to the start of the canal age engineers had demonstrated that water transport was far more efficient than land transport when measured in horse power. It is the same today, even if the horsepower measures output of diesel engines rather than the work that can be done by an ambling quadruped. Emissions from a barge are only a quarter of those from a truck for an equivalent load. Carrying by water makes sense, but it will never happen unless the carriers provide the service the customers want. It has always been a bit unfair to compare what is happening on mainland Europe with the British scene. Continental waterways are far more extensive, have been through huge modernisation programmes and extend over a far wider network.

The great rivers have always been a vital part of the European economy, and in the 1990s two of the most important, the Rhine and the Danube, were linked by a canal, the Rhine–Main–Danube. This provides access to literally thousands of miles of continuous waterway. The canal was built to a suitably grand scale, able to take vessels up to 190m long and 11.45m wide. In other words, any lock could hold half a dozen of the Humber barges at the same time. Standing by one of the busy European waterways is a revelation, with a constant stream of vessels of all kinds, from big barges to push-tows with perhaps half a dozen barges being shoved along. There is still huge scope for improvement – the current aim is to see 11,000 commercial vessels working in the EU. That things have developed as far as they have is the result of a concerted effort at modernisation, both of the waterways themselves and the vessels that use them. What has Britain been doing in the last half century?

A considerable amount of work was done, starting in the 1950s, to improve old navigations, such as the River Lea and the Sheffield and South Yorkshire Navigation. But improved waterways are no use, unless there are carriers who want to use them and an identifiable and viable trade. The most enterprising ideas involved trying to link in directly with that huge European system. It is obviously possible to ship a container down to Rotterdam, load it on a ship, carry it across the North Sea, and unload it onto a barge. A much more interesting idea would be to load the container onto a barge somewhere in the heart of Europe and then unload the same barge in, say, Leeds. BWB had already started a programme of building push-tugs for use with standard 140-tonne compartment boats, when in 1967 they commissioned a study to design a vessel that could link north-east England with Europe. There was already a system called LASH, lighter aboard ship, which was, as the name suggests a way of loading barges onto mother ships. A Danish shipbuilder came up with a new variety BACAT, barge aboard catamaran. The vessel was designed to carry ten of the new compartment boats and three LASH lighters, stowed either on deck or between the twin hulls. The idea was that the vessels would operate between the Humber and continental Europe. There was nothing whatsoever wrong with the system in terms of practicality. It was ultimately to fail through misguided opposition and a failure of political will.

Canals helping to build for the future; barges delivering material to the construction site for the London Olympics.

Construction of the first BACAT began in 1974 and there was a long period of consultation on how the system would work. Everything seemed set for a successful operation. The whole beauty of the system was that it saved on time-consuming and expensive loading and reloading of cargoes, which made economic sense – unless you happened to be a Hull docker who would otherwise have been doing the work. There was an unofficial action group, who made it their task to kill the plan before it even got started. They could not black the BACAT ships themselves, but they could and did threaten to black all BWB vessels in the north-east. The BACAT operators were prepared to fight, but neither BWB nor the Government backed them. The dockers had won, but it was a hollow victory. BACAT might have given a boost to the whole idea of waterways transport, but now its failure confirmed old prejudices. The canal system was antiquated, plagued by disputes and of little value in the twentieth century. An opportunity was lost – and in the end no jobs were saved.

The BACAT story is a sad tale of wasted opportunities, but even if it had succeeded it is by no means certain that there would have been any great increase in freight traffic over the system as a whole. Waterways were, and by many people still are, seen as the product of the eighteenth-century Industrial Revolution. They were best dumped along with the waterwheel and the steam engine or preserved as museum pieces. Global warming has changed the perspective. Numerous schemes have been put forward in recent years, of which the most interesting was the improvement of the Bow Back Rivers to make them a part of the 2012 Olympics site. These backwaters of the Thames had scarcely been used in the last half century and their revival depended on the construction of a new lock, the Prescott Lock. The intention was that a large part of the construction material for the site could arrive by water and afterwards the waterways could be used to remove refuse from the new homes and attract all kinds of leisure activities, from water taxis to floating restaurants. It is just the sort of imaginative idea that deserves to succeed, but there were the inevitable engineering problems. Happily these were overcome, and at the time of writing the first barges were already at work.

One of the greatest changes that has happened in the canal world in the past fifty years has been the way in which attitudes have changed. It has slowly dawned on people that they are national assets that have an important role to play in both urban and rural environments. One of those who tried to change opinion was the architect, writer and photographer Eric de Maré. His work was an inspiration to me, opening my eyes to the visual splendours not just of canals but also of much of the early industrial world as well. In *The Canals of England*, published in 1950, he celebrated the rich variety of canal architecture, but also lamented the way in which we had allowed so much of the system to deteriorate:

> English canals are at their worst in towns. Buildings turn their rears disdainfully, discharging stinking effluents. Local knowledge is needed to pilot craft safely between sunken reefs of tin cans, brickbats and skeletal remains. No one seems to be responsible. Here is British squalor at its most horrific. We must now use canals in towns to form delightful strip parks, because water provides the landscape architect with his most useful material.

There were many parts of the system where things were very little better in the 1970s. I remember being asked to deliver a boat from Bradley Yard on one of the further flung reaches of the Birmingham system, the Wednesbury Oak Loop, down to Hillmorton on the Oxford. I was advised not to start the motor until I was out of the yard because of the rubbish in the water, so I poled out into the middle of the canal as suggested. I had gone no more than a few yards when there was a horrendous crunch and the propeller stopped. I lifted the weed hatch, felt down through the greasy water and began clipping away at a tangle of wire and cloth. It was quite clear that this was a hopeless task. I poled back into the yard. The stern was hoisted clear of the water to reveal an indescribable mess that must once have been a substantial piece of furniture, but was now a tangle of broken springs and cloth. My decision to abandon trying to cut it away with a pair of pliers was more than justified – in the event it took half an hour with an oxyacetylene burner before I could go on my way again. Such were the delights of urban travel some thirty years ago.

Because of the situation described so pithily by de Maré there was a general feeling that if the authorities didn't care why should anyone else? So the canals became dumping grounds for anything at all – on a later occasion I found a canal entirely closed by a dumped Austin Mini sitting right in the middle of a bridge hole. There were some who recognised that urban canals could be very different. As early as 1951 a passenger boat service was started on the Regent's Canal, using the traditional narrow boat *Jason* and the butty *Serpens*. The Regent's Canal was always a special case. Instead of running round the backs of factories it went past the elegant houses of Little Venice, round the edge of Regent's Park and past London Zoo. It took rather more of an act of imagination to persuade people that Birmingham could offer similar delights.

In the late 1960s, new tower blocks were built very close to Farmer's Bridge Junction at the heart of the BCN. Birmingham City Council decided that the new residents deserved something better to look at than the usual set of run-down buildings that bordered the canal. BWB agreed to dredge the former basin of the old Newhall Branch, while the city did something about the surroundings. At that time Birmingham City Council employed a young architect, who actually lived on a narrow boat at nearby Gas Street Basin. He recognised that what the area needed was not the destruction of the old, but its renovation. Here were original eighteenth-century cottages, with all the qualities for which the period is famous. He also appreciated that canals had their own rugged visual language – stone sets and iron bollards for example provide a wonderful contrast in texture, shape and colour. The area also needed a focal point for residents, not just for boaters: what it wanted was a canalside pub. That was added to the mix. The Long Boat was given a modern design, but one that fitted well with the surroundings. The final result was the Cambrian Wharf development that won a Civic Trust award in 1970. Eric de Maré had hoped to see just such a scheme developed to revive the jaded canals in cities: here it was demonstrated in practice and it worked.

In 1968 Sir Frank Price, who had been in charge of Birmingham City Council when the Cambrian Wharf scheme went ahead, was appointed Chairman of BWB. He soon recruited Peter White, the young architect who had worked on the scheme, and it has always been Peter's proud claim that he was the first architect to be employed on canals since Thomas Telford. When, with Mark Baldwin, I edited a book of essays to mark Charles Hadfield's seventy-fifth birthday, I invited Peter to contribute one of the essays. It turned out, though it came as no surprise, that Eric de Maré had also been one of his greatest influences. In his essay he explained his own growing enthusiasm:

> I began to understand how use and wear had imparted their own delightful patina to be added to that of erosion, the weather and the passage of time to create fascinating effects. I was introduced also to a friendly unassuming world, an environment that was charmingly straightforward, sometimes simple, often somewhat neglected and slightly rough at the edges, a world that was frequently and visually arresting.

Peter had identified an important point about the visual appeal of the canals: it was not just about the grand effects, the soaring aqueducts and long flights of locks. Just as important were the details, the gentle curves of stone steps at a lock, worn down by generations of boatmen; the swirling patterns cut into bollards by countless mooring ropes. I first travelled the canals with Peter in the early 1970s and I still have the old blue BWB canal guides that I used. They are liberally spattered with Peter's sketches, often made on the spot as we chugged by, but capturing the essentials. What he spotted, time and again, were the telling details, such as an otherwise unremarkable bridge near Bunbury on the Shropshire Union. He spotted how the engineers had realised that the bridge abutment could be pierced to create a snug, dry resting place for stop planks. It is this appreciation that the small things are at least as important as the big that has been one of the characteristics of his time at BWB.

It was Peter's concern with preserving the essential character of the canals that led him in 1972 to develop a manual *Waterway Environment Handbook* for use by BWB engineers and maintenance staff. The golden rule was – do not do too much. The canal structures have been there a long time and there is little we can do to improve them – but we can very easily ruin them. His approach was soon justified:

> A Civic Trust Award (1972) on the Regent's Canal endorsed this approach by applauding the Board's efforts 'for resolutely refusing to overstifle the canal vernacular'. Conserving our past therefore does not mean spending vast sums – because here we received an Award for doing virtually nothing! Thus a sensitive campaign designed to conserve and facelift the canal fabric was launched, and since 1972 its impact has been quite extensive.

Peter is too modest; without his campaign and the laying down of simple rules we would not have the rich historic fabric that somehow still survives through much of the system. Peter

Birmingham's canals, once neglected, have now become the focus of development.

has never been unaware of the broader picture, nor does he see conservation as being just about the preservation of the past. Changes are bound to come – the important first rule should be that the new should adhere to the same values as the old. There should be the same respect for place, a certain functional simplicity and appropriate use of materials.

Over the years, as the canals have become more popular, and their attractions more widely recognised, so local authorities and developers have started to take an interest. Nowhere has this been more dramatically demonstrated than in city centres, such as Leeds, Manchester and Birmingham. It is an old, often quoted fact that Birmingham has more miles of canal than Venice. It has to be said that this was the only point of superiority. I have always enjoyed going through Birmingham by canal, though in places it was a dispiriting experience, drifting past abandoned warehouses with broken windows and paintwork like flailed skin, with frequent stops to clear rubbish off the prop, whose origin you were generally happier not recognising. For a long time it seemed as if no one cared and that the best was slowly but steadily being destroyed.

When I first came this way the system was full of secret delights, of which Gas Street Basin was perhaps the best of them all. If you were not actually on the canal then you would hardly know it existed. The only road entrance was screened behind a high brick wall, and the only clue that there might be something interesting behind the wall was a red painted wooden hatch. This was a fire point – the fire brigade could open it up and let their hoses down into the canal as a handy water supply. Once inside it was a closed world, of old canopied warehouses surrounding the basin full of traditional narrow boats. Early in the 1970s I wrote about this place:

> Every city needs its Gas Street: odd places, different places, not places done up for tourists, but there giving delight to the curious who take the trouble to look for them. It's a touch of humanity, a place that has developed its character from the people who use it, that's happened almost by accident. Try to improve, plan it and develop it and the whole thing's ruined.

Shortly after those words appeared in print I had a phone call; there were plans to demolish the old warehouses, would I join in the protest – of course I would. Next morning the phone rang again – the bulldozers had already moved in. There was nothing left to argue about. Next time I went back, the basin was exposed to an industrial wasteland that was being used as a wind-blown car park. The mystery had gone and the beauty with it. It has since been redeveloped and not badly done, but one has to think that if the development had been done today, the old would have been respected and adapted and the result would have been far happier.

Thirty years later, Birmingham has been transformed. The canals now lie at the very heart of new developments. It is still not Venice, but it makes a pretty good shot at being Britain's Amsterdam or Bruges. Instead of old factories, important buildings such as the National Indoor Arena and the International Convention Centre look out over the water. Cafés and restaurants line the towpath. I remember all those optimistic architects' drawings of people sitting outside under colourful umbrellas, watching the boats go by, and it all seemed hopelessly unlikely. That was not the British way. But it has happened; that is just what it is like. There has been a price to pay. Something of the original character has been lost, though by no means all. The march of locks up the hillside at Farmer's Bridge is as dramatic as it ever was, and no one really misses the floating dead dogs. On the whole if there has to be change then at least let it be like this, offering an environment that is there to be enjoyed. And it is the canal that makes it special: the passing boats, the wavering reflections of lights at dusk and the sense of the past that still lingers. But I still miss Gas Street.

The newly discovered attraction of a canalside site for anything other than industrial development is something of a mixed blessing. If a home by a canal seems appealing to a family, then it is going to be no less appealing to a developer. This is not necessarily a problem in towns and cities, which have seen some very interesting new buildings over the years, such as the modern metallic houses overlooking the Regent's Canal in Camden Town. Unfortunately most new domestic building in Britain is characterised by mediocrity, a series of dull, unremarkable brick boxes. Rather more worrying is the possible spread of housing out along the line of the canal, much in the way that the brick semis spread out along arterial roads in the 1930s. This is something that needs to be controlled by the waterways authorities and local planning officers. The last thing we need is for canals to be transformed into elongated garden ponds behind rows of dreary houses.

It is always difficult looking into the future. When I first became interested in canals I never foresaw that restoration would proceed at the pace that it has, and when I cast doubt on the possibility of ever reopening some canals I have been delighted to be proved wrong. I am not going to guess what will happen in the years to come, but I will give my opinion for what it is worth of where I think we should be looking. My heart has always been with Tom Rolt and his love of the working boats on the narrow canals, but my head has long since been convinced that those days will never return. But his other argument, that the special character of the canals derives directly from their working past, is as true now as ever. If we lose that then we have lost something unique and irreplaceable. In the final chapter I shall try and identify what those special characteristics are and why they matter. Attempts to bring more freight traffic back to the canal have been going on for decades, with little effect. It may well be that global warning and the looming fuel crisis will succeed where rational argument has largely failed. It will be good for us all. To quote Charles Hadfield: 'The history of British canals continues.'

15

LOOKING TO THE FUTURE

THIS CHAPTER is not intended to offer a balanced view of either the canal system today or of what it should be like in the future. It is a very personal statement, in which I shall try to explain what excites me about the canals, what has brought me back to them time after time and why I believe they are important. It will mostly be concerned with the canals that today are used entirely for leisure and are likely to remain that way into the foreseeable future, but I will begin with a brief word about the commercial waterways.

I first started writing about canals nearly forty years ago, and in all that time there have been plans put forward at regular intervals, promising exciting new developments in carrying freight by water. Sadly, few of the dreams have ever become reality. If anything is going to happen, then it has to be now. There has never been a better time for change, given the unprecedented dangers to the environment that we all face. Whether anything is actually done remains to be seen. When the Government still seems to be able to approve extra runways for international airports and fund extra lanes on motorways one cannot feel wildly optimistic that they are taking environmental issues seriously, but common sense may still occasionally come into the discussions on transport policy.

If there is to be a future for water transport it will have to make use of new technology with no more than passing regard for the historical fabric of the waterways concerned. There is no room for nostalgia: efficiency has to be the key to everything. This was what ruled the thinking of engineers who built the first canals and what they showed was that 'functional' did not have to be another word for 'ugly'. There is evidence of this throughout the system, and it is only time that has given a romantic aura to structures and canal craft. The narrow boats were not designed to be picturesque; they were the best practical answer to a specific problem. They were there to do a job. Intrinsically, they are neither more nor less beautiful than a modern barge. Recently I sat beside one of the busiest waterways in Europe and I had as much pleasure in seeing the continuous traffic of motorised barges and push-tows as I would have done sitting by an English canal and watching a horse-drawn narrow boat drift by. The waterways were built for use, and where they can still be used they should be. I also believe, that left to their own devices, modern engineers are perfectly capable of building as well as their predecessors, as the Falkirk Wheel gracefully testifies.

The story of the remaining canals is very different. Efficiency is not the first priority. A very large proportion of the people who turn to canals for pleasure do not even use boats at all. They stroll along the towpath, go for a drink at a canalside pub and thousands sit on the bank,

A well-used bollard, showing the marks of many   Old paddle gear on the Pocklington Canal has a
mooring lines – and an interested spectator.   sculptural quality.

fishing. I don't know if I am unusual, but I must have drifted past hundreds of anglers over the years, and not once have I ever seen one catch a fish. For me it remains a mysterious and baffling activity. The rest, those who travel by boat, do so, in part at least, to get away from the rush and stress of modern living. If they were in a hurry to get from A to B they wouldn't be choosing a canal boat for the job. There is, however, an argument that things should be made as easy as possible for holiday boaters, and that can lead to problems for those of us whose main interest is in the historic framework of the canals. It can be reasonably argued that people like myself form a quirky minority, which may well be true, but I would argue that the case is not really that straightforward. Consider, for example, the introduction of hydraulic gearing. It's worth looking at it a little more closely, and I had better come clean right at the start: I disliked it from the first, and I have never seen any reason to change my mind.

The whole point of hydraulic gearing is that it is supposed to take the hard work out of boating. Why, in the twenty-first century, should holidaymakers have to wrestle with cast-iron monstrosities designed over two centuries ago? One part of my answer would be to argue that modern is not necessarily the same thing as better. It is an immutable physical law that you get nothing for nothing. Basically, you are still going to be turning a handle in order to lift a heavy sluice gate. With hydraulic gear, each turn is easy, but you have to make a lot of them, far more than you would with conventional gear – and you have just as many turns to wind it down again. With traditional gear, each turn requires a bit more effort, but there are fewer of them. The usual problem with the 'old-fashioned' gear is not that it is inefficient, but that it is sometimes badly maintained. It is like the little girl in the poem: when it is good it is very, very good, but when it is bad it is horrid. Even then, the worst gear can eventually be made to work with enough effort. When hydraulic gear goes wrong there is nothing to be done until an engineer comes along to fix it. A common problem, which everyone who has travelled the canal extensively will have come across, is the slipping hydraulic gear. Round and round goes the handle and eventually the paddle is fully lifted: you let go the handle and it immediately starts whirling back again in the opposite direction. The paddle you have just laboriously raised slides back down. There is no simple mechanical catch to hold it in place as there is on the older gear – a cue for frustration and a lot of bad language. But even if hydraulic gear was brilliant and foolproof, I would still dislike it. The reason has everything to do with what I consider the essential character of our canals.

Over the years, different canal companies developed their own styles of paddle gear. Some was enclosed, like the gear on the Grand Union. Some was partially enclosed, a favourite system on the Trent & Mersey. But mostly it was open, and that's a good part of the pleasure. You can see exactly what is going on and how it works. The lock handle is applied directly to the spindle to operate a simple ratchet gear mechanism. But even at its most basic, there are no ends of variations, and each has its own appeal. On the Bridgwater and Taunton, for example, the ratchet has a counterbalance, consisting of a heavy weight on a chain that runs over a pulley. It looks very Heath Robinson, but it works. Some canal companies opted for very different systems, such as the jack clough of the Leeds & Liverpool. This operates on a lever system, which instead of moving the paddle vertically, swings it across the face of the opening. It is a reminder that the canal system was not developed as a coherent whole, but piecemeal by a variety of different companies, employing different engineers, who had their own ideas of how things should be done. The result is a transport system of great richness and that richness derives from the small details just as much as it does from the grand gestures. Lose this diversity and you lose something that will never be recovered. Does it matter? I believe it does.

We have inherited an extraordinary system, designed to meet the transport needs of the eighteenth century. It is a unique survivor from a time when changes took place in Britain that affected the whole world. The canals of Britain are far more than just any old historic leftover; this was the transport system that was crucial to the development of the Industrial Revolution. This was not some little local event but one of the greatest changes in recorded history. It happened right here in Britain and we should no more think of destroying one of the most important physical survivors of that history than we would think of knocking down a Greek temple. The canal system is not a single building, nor an isolated structure, but a whole complex system. It has a story to tell, and like all the best stories it makes its effect as much by the accumulation of details as it does from the broad sweep of narrative.

There is a powerful argument for considering the canal system as a historical monument of international significance, but saying that does not tell you what to do with it. With ancient buildings many conservationists take the view that all that should be done is to stop time; add nothing and take nothing away, preserve things exactly as they are found. This does not work with a system that has to be used, and used in many very different ways.

In recent years the appreciation of canals has grown, not just among the traditional supporters who echo the famous lines from *The Wind in the Willows*: 'There is nothing – absolutely nothing – half so much worth doing as simply messing about in boats.' A growing number of walkers are discovering the delights of strolling along towpaths. I have done a lot of that myself, and I particularly enjoy walking along the line of abandoned canals, discovering half-forgotten locks and bridges that no longer cross anything and lead nowhere. This has produced a certain conflict of interest. A favourite walk near my home is along the Thames & Severn Canal, down the Golden Valley between Chalford and Sapperton. There are few more beautiful stretches of canal anywhere in the country and there are very many who appreciate it just as it is, wild and overgrown. They see it as a splendid wildlife habitat, a place to walk in peace and quiet, with little to hear beyond the song of birds. They do not want it restored. They do not wish to see it tidied up, nor do they want the peace disturbed by passing motorboats. I have a great deal of sympathy with these views. If restoration does go ahead, their opinions need to be taken into account. It would be intolerable, for example, to introduce hard concrete edgings to the canal, destroying the sense of the canal as being at one with the landscape that it enjoys at present. Equally, if new structures are needed, they have to be sympathetic, using wood and local stone just as the original builders had done. In other words, the canal has to be seen as part of a wider environment, not just as a water-filled trough for boats. So now we have added another criterion to the need to preserve the historic fabric, a concern for the environment.

In the end, consideration of how canals will be treated in the future will depend on where you place the emphasis, on establishing priorities. If we are not discussing commercial freight operations, then I have absolutely no doubt what the first priority should be. Britain's canal system is a unique survivor, which miraculously has preserved its character through more than two centuries of use. If it was a system that had been built as a whole, then you could keep just one bit of it and let that stand for all the rest. But it is not like that. What we see today depends on decisions taken by individuals, who were guided by what needed to be done and the materials that were available to achieve that end.

A very good way of seeing how many different factors contribute to the diversity of the canal scene is to look at the humble bridge. These are such a common feature of all canals that it is easy to take them for granted, yet they represent a major part of the whole construction process. If you travel the whole length of the Trent & Mersey, for example, you will have passed under 213 numbered bridges before you finish your journey. Multiply the numbers up to cover all the canals of Britain and it is clear that over the years literally thousands of canal bridges have been built. Some will be carrying major roads, others quiet country lanes. Many will have been accommodation bridges, built because the canal was cut through the middle of a farmer's fields, and have probably never seen heavier traffic than cows heading to the barn for milking. Whatever the bridge, a good deal of thought went into designing a structure that was suitable for the job.

In general, the best available local material will have been used – stone if quarries could be found near the line, or bricks baked near the site, using clay dug locally. Canal companies were often hard pressed for cash and when finances got really low they found it cheaper to build movable bridges. Some are very simple. The typical lift bridge on the Oxford Canal is no more than a wooden platform, moved by heavy beams set at an angle to act as a counterbalance. Those on the Llangollen are altogether more elaborate. These are bascule bridges, with overhead balance beams. These are very easy to lift, as all you have to do is haul on the chain hanging down from the overhead balance beam, with gravity doing most of the work for you. This assumes that no one has vandalised the bridge and removed the chain, in which case there is nothing to be done but to practice your weight lifting skills and push the end of the bridge upwards. At least that demonstrates very conclusively just how valuable the working bascule bridge is in saving effort. The other alternative is to have a swing bridge, moving horizontally. Opening these is generally supposed to be very simple, leaning against a balance beam much as one would when opening a lock gate. It is here that the problem with the moveable bridge is likely to become all too apparent. A heavy bridge pivoted at one end is always liable to droop, and bracing is needed to keep it in the horizontal. In time, however, the platform will inevitably begin to sag. When that happens, the end will stick. When I first travelled the Peak Forest Canal shortly after restoration, it seemed that in a long procession of swing bridges there wasn't one that didn't require a mighty effort from two pretty hefty blokes to shift it. The canal company may have saved money during construction, but it was the canal users who paid the price.

Bridges are the punctuation marks of canal passages: a lift bridge on the Llangollen Canal.

The most interesting bridges are turnover bridges, where the towpath changes side. It is easy for the horse to walk over a bridge, but it is not going to drag the boat over as well. The easy answer is to make a long approach ramp to the bridge and then add a second ramp on the opposite bank, doubling back, so that the horse now walks back under the bridge once it has got across to the opposite bank. A far more elegant solution can be seen on the Macclesfield Canal, where there is a conventional ramp on one bank, but then it curls up to pass under the arch on the opposite side. These are often known as 'snake' bridges, because of the coil, but in many ways they are even more reminiscent of the spiral shell of a snail. The introduction of iron offered the possibility of an even simpler solution. When working in stone or brick you have to construct a complete arch, but with iron you can leave a gap. So-called 'split bridges' aren't really split at all. They are simply two short cantilevers, rather like diving boards, that don't quite meet in the middle. The gap in between is just big enough to allow the towrope to pass through. The device is equally useful where bridges cross the tail of a lock.

Today the material of choice for most bridges would be concrete, and there is no reason why it should not be used imaginatively. Sadly, it rarely is, as any trip down a motorway will confirm. No doubt, if it had been readily available in the days of Brindley or Telford they would have welcomed it as easy to use, practical and economic. But it wasn't, so they used what was available. It is our good fortune that the materials they chose have improved with age. Bricks mellow, stone loses its harsh edges and the passing of time brings in changes. The endless procession of tow ropes has cut grooves and worn away corners so that many bridges have developed sculptural details that add to the visual pleasures of the scene. Concrete neither mellows nor develops rich tones: it simply stains. The best that can happen is that it will be well maintained and regularly treated so that it keeps its original crispness and wears a pristine white coat. The old canal bridges seem to have slid back into the landscape, reuniting with the

Good design is timeless: a simple cast-iron bridge number.

Stone steps beside a bridge on the Rochdale Canal and drystone walling make a satisfactory study in contrasting textures.

earthy materials from which they were built. You cannot create such effects; only rely on time to produce them for you.

It is possible to go on listing all kinds of structures that add to the charm and interest of canal travel. All of them were put there to serve a practical purpose, not as mere decoration. Mileposts were essential markers and they come in a rich variety of styles, whether elegantly carved in stone on the Lancaster or crisply cast in iron on the Shropshire Union. What could be more mundane than an overspill weir, basically no more than a drain to allow water to overflow from the canal down to a nearby river or stream? Yet they can be attractive and varied. Those on the Macclesfield are beautifully constructed with a surface of stone setts but quite conventional in form, while Brindley often preferred to build circular weirs, like giant plugholes. Bollards are simply posts to tie a rope round, but time and use has often turned them into pieces of abstract sculpture. One could go on, but there is no need to labour the point. The visual delight of canals is made up of different elements that all combine to produce the effect. There is a pleasure in seeing buildings and structures ideally fitted for their purpose, and the use of local materials ensures a harmony with the surrounding landscape. These are all factors that can be reproduced today. But there is one element that cannot – the patina of age. But if we build as well as our predecessors, then time will take care of that as well.

We can save the old if we make it a priority, but there will always be a demand for change, not least during restoration. Do you try and match the old, producing some sort of historical pastiche or do you go for something daringly modern? Both approaches can work. During the restoration of the Huddersfield Canal, there was a major problem to tackle at Slaithwaite. A road bridge had been dropped and a large part of the canal had been covered in.

The solution was to move a lock to lower the canal, rather than raise the bridge and construct a box culvert out of concrete and sheet piling. The result was harsh and uncompromising and, if left in that state, even the most ardent proponent of the function-is-all theory would have been hard pushed to describe the result as anything but ugly. But that was not the end of work. The structure was given a cosmetic stone cladding, very much in keeping with the dominant stone themes of this typical Pennine mill town. Purists might describe it as dishonest, but it works. Castlefield Basin in the heart of Manchester has been transformed as part of the restoration of the Rochdale Canal. The most recent addition is a footbridge, with the walkway suspended from arches. The result is elegant and uncompromisingly modern. It works perfectly here, because there is already a huge mixture of styles on show, from the simple early canal bridge to the castellated Victorian railway viaduct. It all goes to show that there is no single answer to restoration problems; all that matters is quality and a sensitive response to what already exists.

There is one worrying development of recent years. The early canal promotion for the Trent & Mersey had enthusiastically recommended 'a lawn terminated by water'. This was intended to appease landowners who might object to having a canal near their home, but the same message has now reached the ears of property developers. The success of restoration schemes has made the idea of living next to a canal seem very enticing. New housing was discussed in the last chapter, with words of caution. And the same need for caution applies to conversions. There is nothing intrinsically wrong with converting old canalside buildings to new uses, and indeed it is often the only route to preservation. There are many outstanding examples, such as Saltaire on the Leeds & Liverpool. The closure of the great mills of Titus Salt brought an air of glum dereliction to this important example of an industrial village. All that has changed thanks to a sensitive restoration programme and imaginative use for the old indus-trial buildings. The important word here is 'sensitive'. What can be done to preserve what we already have and to protect it from insensitive and inappropriate developments? One possible solution is to create a conservation corridor, say a quarter of a mile wide, along the designated canals. In effect this would create a Canal National Park. It may sound unrealistic, but the total area for even 1,000 miles of canal would only be 250 square miles, which is quite modest compared with the Lake District National Park, which covers 885 square miles. And there is a precedent. The Norfolk Broads have been similarly protected and have become a National Park in all but name. Such a scheme would neither prevent progress nor stop all developments, merely ensure that anything that is done conforms to recognised guide lines. It is an idea that it is at least worth serious consideration.

Whatever the future of Britain's canals might be, they have enjoyed a glorious past. For almost a century they were the vital arteries of the Industrial Revolution. They are memorials to the ingenuity of the engineers who designed them, some of whom became rich and famous as a result. And they also commemorate the immense hard work of the great army of navvies, none of whom became either rich or famous. Our modern transport system of roads and airports has provided us with far more efficient ways of getting about, but they will never be anything more than ugly scars on the landscape. No doubt, when they were brand new, there were many who looked on the canals as little more than ugly gashes across the face of the land. But, because of how they were built, those gashes have healed, and the canals have mellowed so that today they seem no more obtrusive than a river meandering through a meadow. They are there for us to use and enjoy and, above all, to preserve.

GAZETTEER

THE FOLLOWING list represents a selection of the most interesting sites to be found on Britain's canals. The map references are to Ordnance Survey Landranger maps. The first number gives the number of the map, the second group of digits provides an accurate reference for locating the site. For example, the first entry for Monkey Marsh lock tells you that it is to be found on Landranger 174, which is the map covering Newbury and Wantage. The following six digits use the standard Ordnance Survey reference system – there is an explanation of how to use this on every map. Where a site covers a large area, such as a flight of locks, the four-digit number indicates the location to within 1km.

LOCKS

Caledonian
> Neptune's Staircase, Banavie: 41/1177
> Five-lock staircase, Fort Augustus: 34/3709

Grand Union
> Foxton staircase: 141/6989
> Hatton 21: 151/2466
> Watford staircase: 152/5968

Kennet Navigation
> Aldermaston lock: 175/601672
> Monkey Marsh turf lock: 174/525663

Kennet & Avon
> Caen Hill flight, Devizes: 173/9861

Leeds & Liverpool
> Bingley Five Rise: 104/1039

Macclesfield
> Bosley: 118/9065

Monmouth
> Fourteen Locks: 171/269892

Staffs & Worcester
> The Bratch: 139/865933

Worcester & Birmingham
> Tardebigge 30: 139/988689

AQUEDUCTS

Bridgewater
      Barton swing: 109/767976
Kennet & Avon
      Dundas: 172/785625
Lancaster
      Lune: 97/484639
Llangollen
      Chirk: 126/287371
      Pontcysyllte: 117/271420
Peak Forest
      Marple: 109/955901
Shrewsbury
      Longdon-on-Tern: 127/617157
Stratford
      Edstone: 151/162609
Union, Scotland
      Almond: 65/105706
      Avon: 65/967758

TUNNELS

BCN
      Dudley (north): 139/937917
      Netherton (south): 139/953883
Grand Union
      Blisworth (south): 152/739503
Huddersfield
      Standedge, Tunnel End: 110/040120

Thames & Severn
      Sapperton, Coates: 163/966005
Trent & Mersey
      Harecastle (north): 118/838542

EARTHWORKS

Barnsley
      Royston cutting: 111/3613
Birmingham
      Smethwick cutting: 139/0189
Bridgewater
      Bollin embankment: 109/7287
Caledonian
      Laggan cutting: 34/2997
Grand Union
      Tring cutting: 165/9412
Leeds & Liverpool
      Burnley Mile embankment: 103/8432
Shropshire Union
      Grub Street cutting: 127/7824
      Shebdon embankment: 127/7426
      Shelmore embankment: 127/8021
      Woodseaves cutting: 127/6930

TRAMWAY CONNECTIONS

Brecon & Abergavenny
>   Llanfoist: 161/285130
Caldon
>   Froghall Basin: 119/0247
Cromford
>   Cromford & High Peak Railway Wharf: 119/314557
Peak Forest
>   Buxworth Basin: 110/025822
>   Whaley Bridge Interchange: 110/013817

LIFTS AND INCLINED PLANES

Bude
>   Marhamchurch incline: 190/2103
Grand Union
>   Foxton incline: 141/6989
Grand Western
>   Nynehead lift: 181/144218
>   Wellisford incline: 181/102217
Shropshire
>   Hay incline: 127/695028
Tavistock
>   Morwellham incline: 201/445698
Union (Scotland)
>   Falkirk Wheel: 65/8580
Weaver
>   Anderton lift: 118/647753

PUMPING STATIONS

Cromford
>   Leawood: 119/3155
Kennet & Avon
>   Claverton: 172/791643
>   Crofton: 174/262623

WATERWAYS MUSEUMS

Map references have not been supplied for this section, as the museums are well signposted and easy to
   locate. Web sites have been added as an aid to finding out opening hours.

Boat Museum, Ellesmere Port, Cheshire: www.thewaterwaystrustorg.uk/museums

Canal Museum, Foxton, Leicestershire: www.fpt.org.uk

The Canal Museum, Stoke Bruerne, Northamptonshire: www.thewaterwaystrust.org.uk/museums

Galton Valley Canal Heritage Centre, Brasshouse Lane, Smethwick, West Midlands:
   www.laws.sandwell.gov.uk

Canal Museum and Boats, Linlithgow, West Lothian: www.lucs.org.uk

London Canal Museum, New Wharf Road, London N1: www.canalmuseum.org.uk

National Waterways Museum, The Docks, Gloucester: www.nwm.org.uk

Powysland Museum and Canal Centre, Welshpool, Powys: www.powys.gov.uk

INDEX